Mathematics Workbook

AGE 8–10

Mental Arithmetic

David E Hanson

Answers to all the questions in this book can be found in the pull-out section in the middle.

Introduction

- **Arithmetic**: the branch of mathematics dealing with the properties and manipulation of numbers.
- **Mental arithmetic**: arithmetical calculations performed in the mind, without writing figures or using a calculator.

This book consists of 50 tests, each test:

- is printed in two columns on one page
- consists of 20 questions
- has a total of 20 marks.

The tests:

- cover a full range of arithmetic skills
- follow the same general pattern
- feature a gradual increase in difficulty
- are designed to:
  - provide practice in recalling number facts and procedures
  - facilitate the identification of weak areas
  - encourage working quickly and accurately
  - facilitate the monitoring of progress
  - build pupil confidence.

Odd-numbered tests start with 10 straightforward questions which act as reminders of 10 basic strategies. At first there are hints about possible ways forward for each question but these become shorter and less descriptive as the book progresses.

The tests can be used in two main ways:

- Complete the test, as quickly as possible, recording the time taken.
- Do as much as possible in a fixed time.

All the questions:

- are to be tackled entirely in the head, without
  - doing any working
  - any measuring instruments
  - a calculator.
- require a single response only
- are answered on the line to the right or underneath of the questions; units are given where appropriate.

All questions require a degree of mental activity and there are no questions such as 'name this shape'.

Answers to the questions can be pulled out of the middle of the book.

Where appropriate, answers involving fractions should be given in their simplest form.

Test 1

Final score

$\overline{20}$ ____ %

For all of the questions in this test, do the calculation entirely in your head with no written 'working' and just write down the answer.

In questions 1 to 10 you are reminded of 10 useful strategies which may help you in later questions.

1 3×19 __________

* Try 3×20 and then subtract 3

2 £12.80 ÷ 4 £__________

* Divide by 2 and then by 2 again

3 27×5 __________

* Multiply by 10 then divide by 2

4 $53 - 19$ __________

* Subtract 20 then add 1

5 $102 \div 6$ __________

* Divide by 2, then by 3

6 0.9×8 __________

* You know that $9 \times 8 = 72$

7 $\frac{3}{4}$ of 40 __________

* Find $\frac{1}{4}$ then multiply by 3

8 $15 + 23 + 35$ __________

* $15 + 35 = 50$ first

9 1.9×2 __________

* The result must be about 4

10 17×12 __________

* The units digit of the result is 4

11 Jennie has 23 sweets in her bag. She eats 8 sweets and gives 5 to her brother. How many sweets does Jennie have left?

____

12 There are 17 frogs on a lily pad. 9 jump off and 7 climb on. How many frogs are now on the lily pad?

____

13 What is the cost of 4 packets of crisps costing 39 pence each?

£__________

14 Sandra left school at 15:45 and walked for 37 minutes to reach home. At what time did she get home? ________ : ________

15 What is the smallest number, greater than 20, which divides exactly by 6? __________

16 Given that $204 \div 12 = 17$, what is $204 \div 6$?

____

17 By how much is 3×12 greater than 2×13?

____

18 The diagram shows a rug with a pattern of rectangles.

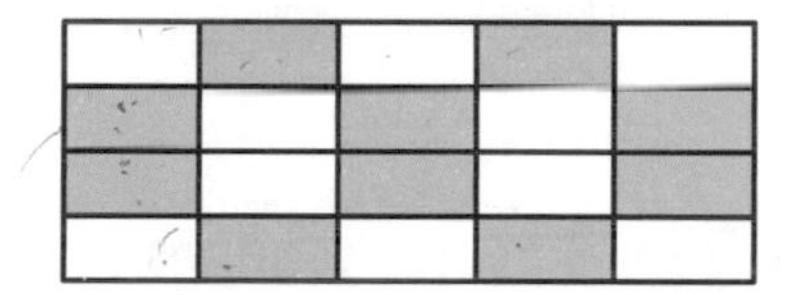

What percentage of the rectangles is blue?

__________ %

19 Shonagh is thinking of a number and gives the following clues:

"My number is:

- less than 20
- a multiple of 4
- 1 more than a multiple of 5"

What number is Shonagh thinking of?

____

20 On this tower of bricks, the number on each brick is the sum of the numbers on the two bricks supporting it.

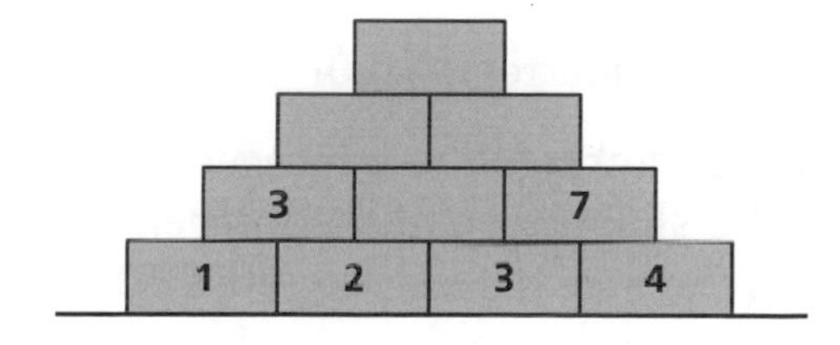

What number is on the top brick? __________

Test 2

Final score

____ / 20 ____ %

For all of the questions in this test, do the calculation entirely in your head with no written 'working' and just write down the answer.

1 Write in figures the number which is thirteen greater than two hundred and two.

____

2 Christmas cards are sold in packs of 10 and cost £1.50 per pack. Guy will need 70 cards for his friends. What will be the total cost?

£__________

3 Find the value of $3^2 - 2^2$ __________

4 What is the perimeter of a rectangle measuring 12 cm by 4 cm? __________ cm

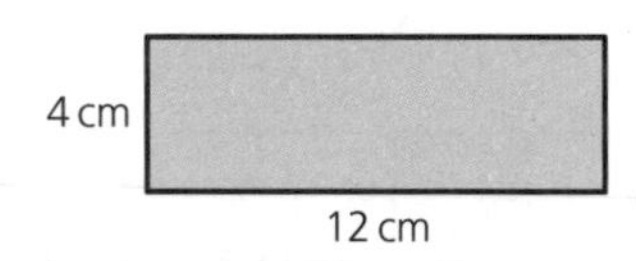

5 A pudding for 6 people requires 300 ml of milk. How many millilitres of milk are required to make the same pudding for 2 people? __________ ml

6 A sweatshirt normally costs £24.00

The price is reduced by a third in a sale. What is the sale price? £__________

7 Joanne won a running race with a time of 3 minutes and 52 seconds. Sophie finished 13 seconds behind Joanne. What was Sophie's time? __________ minutes __________ seconds

8 What is $2 \times 3 \times 4$? __________

9 Robin buys 3 ice lollies costing 99p each. How much change does he receive from a £5 note?

£__________

10 What is a quarter of 120? __________

11 Sean is thinking of two integers (whole numbers). The sum of the integers is 20 and the difference between them is 10

What is the smaller number? __________

12 Ahmed has the three number cards below.

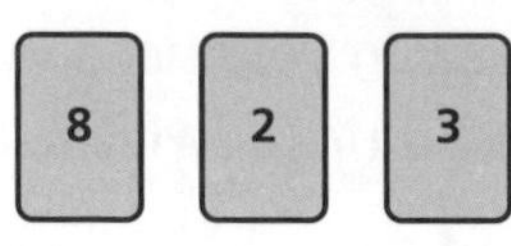

What is the smallest even number he can make by placing all three cards side by side?

____

13 The Venn diagram shows some information about children in Year 5

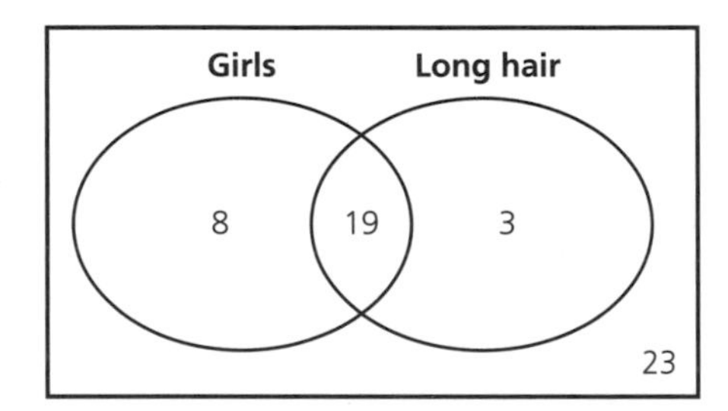

How many girls are there in Year 5?

____

14 The mode of six numbers is 4 and five of the numbers are 3, 4, 5, 3 and 4

What is the sixth number? __________

15 Which number between 100 and 104 is divisible exactly by 3? __________

16 What is the smallest number of British coins needed to pay exactly 47 pence? __________

17 Holly has 20 centimetre cubes. 6 are blue and half of the others are red. The rest are green. How many cubes are green? __________

18 How many of the following calculations give a result with a units digit 5? __________

$8 \times 5 \quad 25 \div 5 \quad 86 - 41 \quad 56 + 109$

19 Tara's birthday is 14 June and Angela's birthday is 20 June. This year Tara's birthday is on a Saturday. On which day of the week is Angela's birthday? __________

20 What is the sum of the first four even numbers? __________

Test 3

Final score

$\overline{20}$ ___ %

For all of the questions in this test, do the calculation entirely in your head with no written 'working' and just write down the answer.

In questions 1 to 10 you are reminded of 10 useful strategies which may help you in later questions.

1 2×29 __________

✱ Try 2×30 and then subtract 2

2 £16.20 ÷ 4 £__________

✱ Divide by 2 and then by 2 again

3 17×5 __________

✱ Multiply by 10 then divide by 2

4 $45 - 29$ __________

✱ Subtract 30 then add 1

5 $96 \div 12$ __________

✱ Divide by 4, then by 3

6 0.7×9 __________

✱ You know that $7 \times 9 = 63$

7 $\frac{3}{5}$ of 60 __________

✱ Find $\frac{1}{5}$ then multiply by 3

8 $17 + 21 + 33$ __________

✱ $17 + 33 = 50$ first

9 2.1×4 __________

✱ The result must be about 8

10 23×12 __________

✱ The units digit of the result is 6

11 Linda has 40 sweets in her bag. She eats 9 sweets and gives 13 to her brother. How many sweets does Linda have left?

____

12 There are 26 butterflies on a bush. 7 fly away and 11 land on the bush. How many butterflies are now on the bush?

____

13 What is the cost of 5 packets of crisps costing 49 pence each?

£__________

14 Kiera left school at 3.30 p.m. and walked for 48 minutes to reach home. At what time did she get home? __________

15 What is the smallest number, greater than 20, which divides exactly by 7? __________

16 Given that $200 \div 8 = 25$, what is $2000 \div 8$?

____

17 By how much is 4×12 greater than 2×14?

____

18 The diagram shows a rug with a pattern of rectangles.

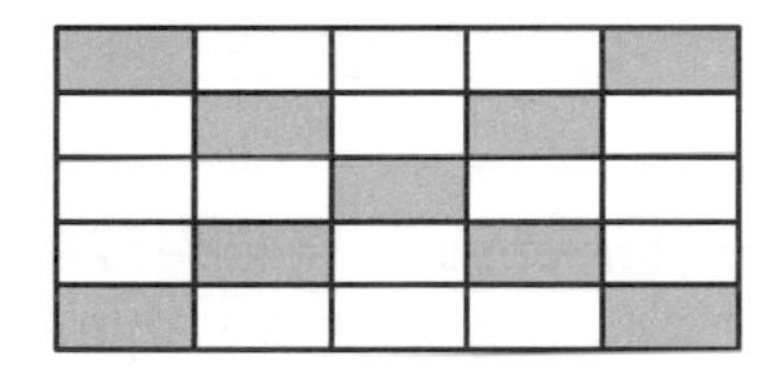

What percentage of the rectangles is green?

__________%

19 Sheena is thinking of a number and gives the following clues:

"My number is:

- less than 10
- prime
- 1 less than a square number"

What number is Sheena thinking of? __________

20 On this tower of bricks, the number on each brick is the sum of the numbers on the two bricks supporting it.

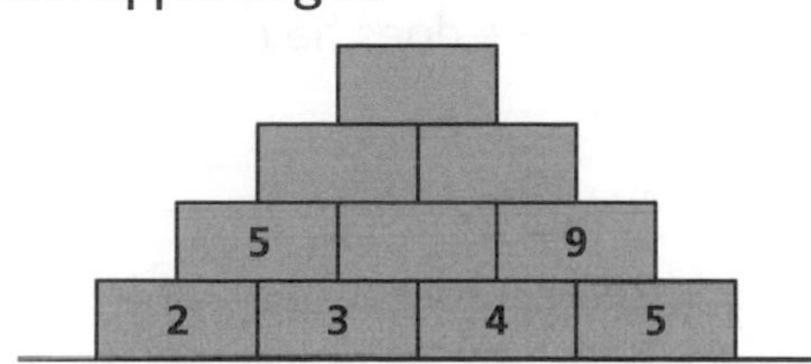

What number is on the top brick? __________

Test 4

Final score

$\overline{20}$ ___ %

For all of the questions in this test, do the calculation entirely in your head with no written 'working' and just write down the answer.

1 Write in figures the number which is twenty-four greater than one hundred and eight.

____

2 Postcards are sold in packs of 5 and cost £1.45 per pack. Liam needs 20 cards to send to his friends. What will be the total cost? £__________

3 Find the value of $4^2 - 2^2$ __________

4 What is the perimeter of a rectangle measuring 11 cm by 5 cm? __________ cm

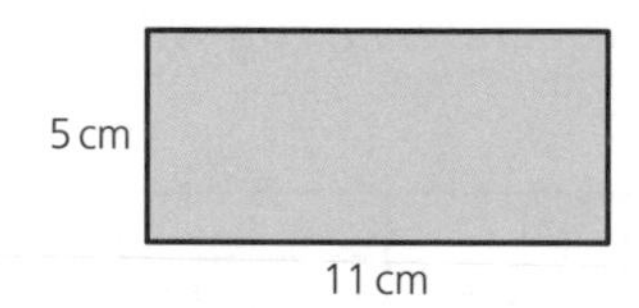

5 A pudding for 4 people requires 200 ml of milk. How many millilitres of milk are required to make the same pudding for 6 people? __________ ml

6 A sweater normally costs £30.00

The price is reduced by a third in a sale. What is the sale price? £__________

7 Tanya won a running race with a time of 2 minutes and 15 seconds. Shanna finished 13 seconds behind Tanya. What was Shanna's time?

__________ minutes __________ seconds

8 What is $3 \times 4 \times 5$? __________

9 Louis buys 3 ice-creams costing £1.10 each. How much change does he receive from a £5 note? £__________

10 What is a quarter of 280? __________

11 Jamie is thinking of two integers (whole numbers). The sum of the integers is 12 and the difference between them is 6

What is the smaller number? __________

12 Brad has the three number cards below.

What is the smallest odd number he can make by placing all three cards side by side?

____

13 The Venn diagram shows some information about children in Year 5

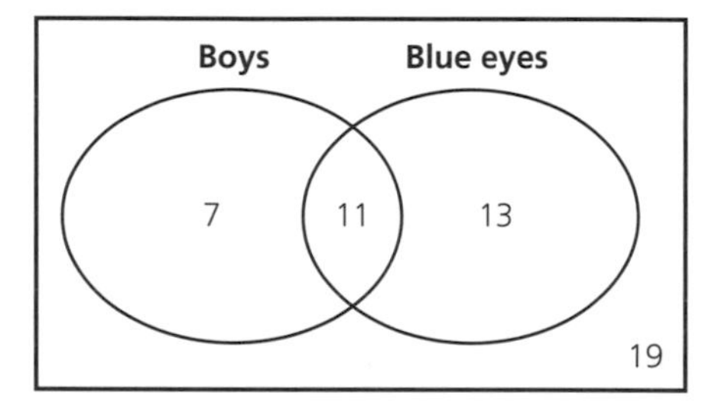

How many children in Year 5 have blue eyes?

____

14 The mode of five numbers is 6 and four of the numbers are 6, 8, 4 and 9

What is the fifth number? __________

15 Which number between 95 and 100 is divisible exactly by 6? __________

16 What is the smallest number of British coins needed to pay exactly 39 pence? __________

17 Stefanie has 40 buttons. 14 are blue and half of the others are red. The rest are green. How many buttons are green? __________

18 How many of the following calculations give a result with a units digit 3? __________

6×9 $21 \div 7$ $41 - 28$ $67 + 76$

19 On 31 December, Louisa's age was recorded as 9 years 11 months. In which month is her birthday?

____

20 What is the sum of the first four odd numbers?

____

Test 5

Final score

$\overline{20}$ ___ %

For all of the questions in this test, do the calculation entirely in your head with no written 'working' and just write down the answer.

In questions 1 to 10 you are reminded of 10 useful strategies which may help you in later questions.

1 3×39 ________

* Try 3×40 and then subtract 3

2 £6.50 × 4 £________

* Double and then double again

3 $240 \div 5$ ________

* Divide by 10 then multiply by 2

4 $45 + 29$ ________

* Add 30 then subtract 1

5 $182 \div 14$ ________

* Divide by 2, then by 7

6 50×7 ________

* You know that $5 \times 7 = 35$

7 $\frac{3}{4}$ of 80 ________

* Find $\frac{1}{4}$ then multiply by 3

8 $29 + 14 + 11$ ________

* $29 + 11 = 40$ first

9 19×7 ________

* The result must be about 140

10 17×11 ________

* The units digit of the result is 7

11 Ben has 30 marbles in his pocket. He loses 7 marbles to Sam and loses 15 to Tom. How many marbles does Ben have now?

____

12 There are 24 swallows on a telephone wire. 5 leave and 16 land on the wire. How many swallows are now on the wire?

____

13 What is the cost of 7 chocolate bars priced at 61 pence each?

£________

14 Karen reached home at 16:20 after walking for 35 minutes on her way back from school. At what time did she leave school?

________:________

15 What is the smallest number, greater than 30, which divides exactly by 9?

____

16 Given that $400 \div 8 = 50$, what is $400 \div 16$?

____

17 By how much is 5×13 greater than 3×15?

____

18 The diagram shows a rug with a pattern of rectangles.

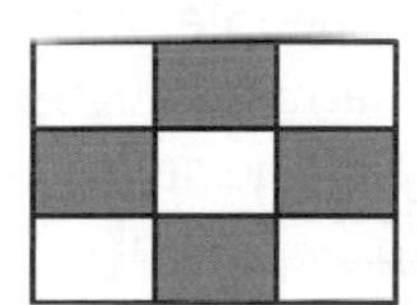

What fraction of the rectangles is red?

____

19 Clare is thinking of a number and gives the following clues:

"My number is:

- between 10 and 20
- a multiple of 3
- 1 less than a square number."

What number is Clare thinking of? ________

20 On this tower of bricks, the number on each brick is the sum of the numbers on the two bricks supporting it.

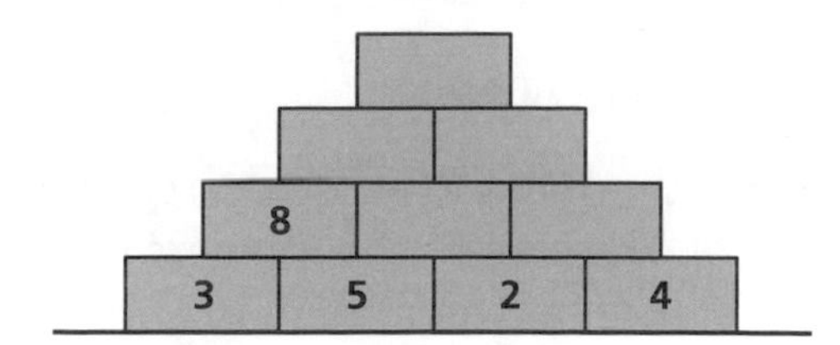

What number is on the top brick? ________

Test 6

Final score

$\overline{20}$ ___ %

For all of the questions in this test, do the calculation entirely in your head with no written 'working' and just write down the answer.

1 Write in figures the number which is thirty-seven less than fifty-five. ________

2 Birthday candles are sold in packs of 6 and cost £1.10 per pack. Pippa needs 42 candles for her mother's birthday cake. What will be the total cost? £________

3 Find the value of $3^2 + 1^2$ ________

4 What is the perimeter of a rectangle measuring 13 cm by 7 cm? ________ cm

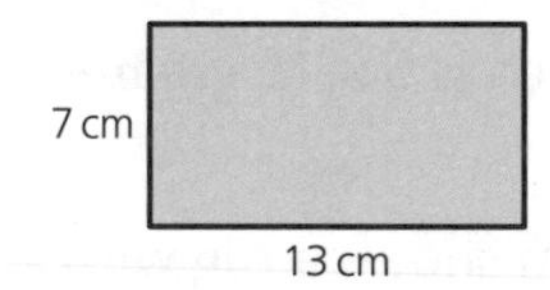

5 A pudding for 6 people requires 240 ml of milk. How many millilitres of milk are required to make the same pudding for 9 people? ________ ml

6 A sweater normally costs £40.00

The price is reduced by a quarter in a sale. What is the sale price? £________

7 Terry won a running race with a time of 2 minutes and 47 seconds. Chris finished 18 seconds behind Terry. What was Chris's time?

________ minutes ________ seconds

8 What is $4 \times 5 \times 6$? ________

9 Max buys 4 ice-lollies costing 90p each. How much change does he receive from a £5 note?

£________

10 What is a third of 180? ________

11 Charlie is thinking of two integers (whole numbers). The sum of the integers is 20 and the difference between them is 4

What is the smaller number? ________

12 Elaine has the four number cards below.

What is the largest even number she can make by placing all four cards side by side?

____

13 The Venn diagram shows some information about children in a club.

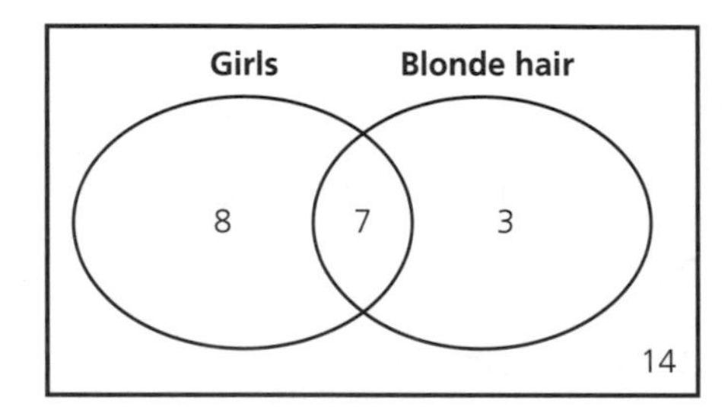

How many boys are in the club? ________

14 The mode of seven numbers is 4 and six of the numbers are 5, 4, 5, 6, 4 and 6

What is the seventh number? ________

15 Which number between 70 and 75 is divisible exactly by 9? ________

16 What is the smallest number of British coins needed to pay exactly 67 pence? ________

17 Justyna has 50 centimetre cubes. 20 are blue and a third of the others are red. The rest are green. How many cubes are green?

____

18 How many of the following calculations give a result with a units digit 4? ________

4×11 $28 \div 7$ $41 - 7$ $22 + 106$

19 Linda's birthday is 15 May and Tom's is 20 May. Linda's birthday is on Tuesday. On which day is Tom's birthday?

____

20 What is the sum of the first four prime numbers?

____

Test 7

Final score

$\overline{20}$ ___ %

For all of the questions in this test, do the calculation entirely in your head with no written 'working' and just write down the answer.

In questions 1 to 10 you are reminded of 10 useful strategies which may help you in later questions.

1 6×49 ___________

✱ Try 6×50 and then subtract 6

2 £256 ÷ 8 £___________

✱ Divide by 2, by 2 again and then by 2 again

3 23×5 ___________

✱ Multiply by 10 then divide by 2

4 $71 - 48$ ___________

✱ Subtract 50 then add 2

5 $165 \div 15$ ___________

✱ Divide by 3, then by 5

6 1.2×7 ___________

✱ You know that $12 \times 7 = 84$

7 $\frac{2}{3}$ of 45 ___________

✱ Find $\frac{1}{3}$ then multiply by 2

8 $34 + 29 + 16$ ___________

✱ $34 + 16 = 50$ first

9 3.9×3 ___________

✱ The result must be about 12

10 19×9 ___________

✱ The units digit of the result is 1

11 Katy has 18 mints in her bag. She eats 4 mints and gives 4 to her brother. How many mints does Katy have left?

____

12 There are 6 children on a roundabout. 2 jump on and 3 fall off. How many children are now on the roundabout?

____

13 What is the cost of 9 packets of crisps costing 59 pence each?

£___________

14 Kirsten left home at 10.55 a.m. and reached the shop at 11.23 a.m. How long did the journey take her?

___________ minutes

15 What is the largest number, less than 50, which divides exactly by 7?

____

16 Given that $360 \div 4 = 90$, what is $360 \div 8$?

____

17 By how much is 8×15 greater than 5×18?

____

18 The diagram shows a pattern of rectangles.

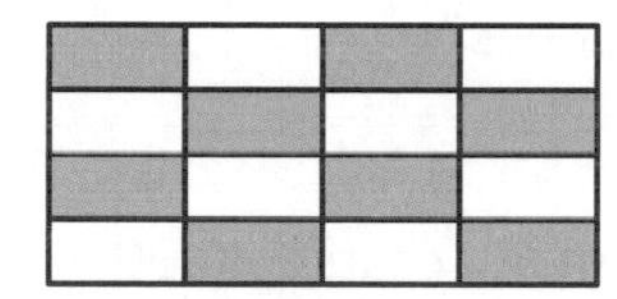

What percentage of the rectangles is purple?

___________ %

19 Stacey is thinking of a number and gives the following clues:

"My number is:

- between 20 and 30
- prime
- 1 more than a multiple of 7"

What number is Stacey thinking of? ___________

20 On this tower of bricks, the number on each brick is the sum of the numbers on the two bricks supporting it.

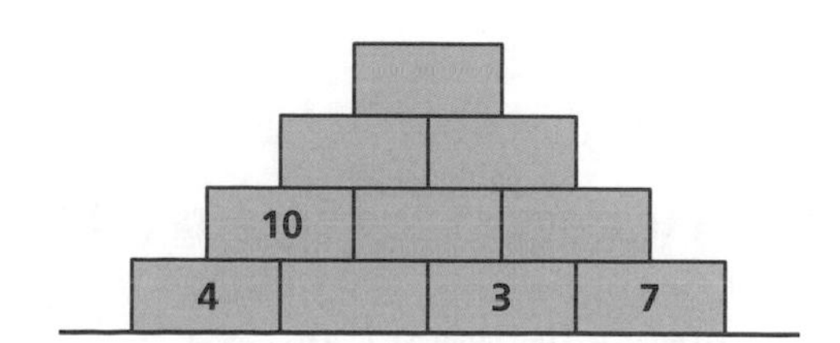

What number is on the top brick? ___________

Test 8

Final score

$\overline{}$
20 ____ %

For all of the questions in this test, do the calculation entirely in your head with no written 'working' and just write down the answer.

1 Write in figures the number which is thirty-nine greater than seventy-nine. __________

2 Stickers are sold in packs of 6 and cost £2.50 per pack. Brad has 66 stickers. How much has Brad spent on stickers? £__________

3 Find the value of $5^2 + 3^2$ __________

4 What is the perimeter of a rectangle measuring 17 cm by 3 cm? __________ cm

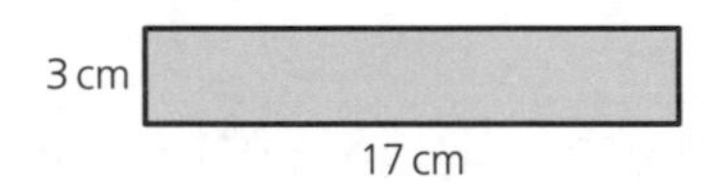

5 A recipe for 6 scones requires 400 g of flour. How many grams of flour are required to make 12 scones?

__________ g

6 A shirt normally costs £20.00

The price is reduced by a 25% in a sale. What is the sale price?

£__________

7 Georgia came second in a running race with a time of 3 minutes and 4 seconds. Fay's winning time was 11 seconds faster. What was Fay's time?

________ minutes ________ seconds

8 What is $3 \times 6 \times 4$? __________

9 Mr Smith buys 2 cups of tea costing 95p each. How much change does he receive from a £5 note?

£__________

10 What is a third of 600? __________

11 Robbie is thinking of two integers (whole numbers). The sum of the integers is 13 and the difference between them is 1

What is the smaller number? __________

12 Ali has the five number cards below.

What is the smallest 4-digit even number he can make by placing four of the cards side by side? __________

13 The Venn diagram shows some information about children in Year 5

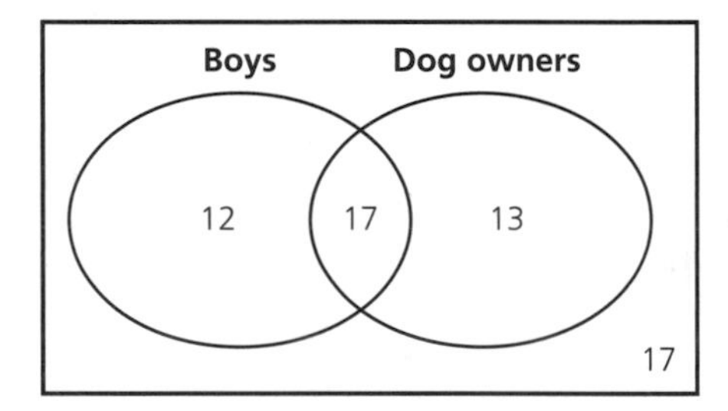

How many children in Year 5 do not own dogs? __________

14 Tim scored the following marks in ten tests.

6 5 9 8 7 4 6 9 7 6

What mark was the mode? __________

15 Which number between 95 and 99 is divisible exactly by 4? __________

16 What is the smallest number of British coins needed to pay exactly 65 pence? __________

17 Natalie has 40 sweets. 12 are mints and half of the others are toffees. The rest are chocolates. How many chocolates are there?

____

18 How many of the following calculations give a result with a units digit 5? __________

14×5 $60 \div 4$ $73 - 48$ $67 + 48$

19 On 31 March, Charlotte's age was recorded as 9 years 8 months. In which month is her birthday? __________

20 What is the sum of the numbers 10 to 12 inclusive?

____

Test 9

Final score

/20 ____ %

For all of the questions in this test, do the calculation entirely in your head with no written 'working' and just write down the answer.

In questions 1 to 10 you are reminded of 10 useful strategies which may help you in later questions.

1 3×49 ________

✱ Try 3×50 and then subtract 3

2 £8.20 $\times$ 4 ________

✱ Double and then double again

3 $220 \div 5$ ________

✱ Divide by 10 then multiply by 2

4 $63 + 39$ ________

✱ Add 40 then subtract 1

5 $252 \div 12$ ________

✱ Divide by 3, then by 4

6 0.8×5 ________

✱ You know that $8 \times 5 = 40$

7 $\frac{2}{5}$ of 70 ________

✱ Find $\frac{1}{5}$ then multiply by 2

8 $45 + 37 + 25$ ________

✱ $45 + 25 = 70$ first

9 11×19 ________

✱ The result must be about 200

10 13×7 ________

✱ The units digit of the result is 1

11 James has 50 marbles in his pocket. He loses 13 marbles to Sarah and loses 9 to Tina. How many marbles does James have now?

____

12 There are 9 beetles hiding under a stone. 6 more hide and then 14 leave. How many beetles are now under the stone?

____

13 What is the cost of 8 chocolate bars priced at 79 pence each?

£________

14 Kevin reached home at 21:10 after walking for 23 minutes on his way back from the disco. At what time did he leave the disco?

________:________

15 What is the smallest number, greater than 50, which divides exactly by 11? ________

16 Given that $169 \div 13 = 13$, what is 13×13?

____

17 By how much is 4×13 greater than 3×14?

____

18 The diagram shows a regular polygon.

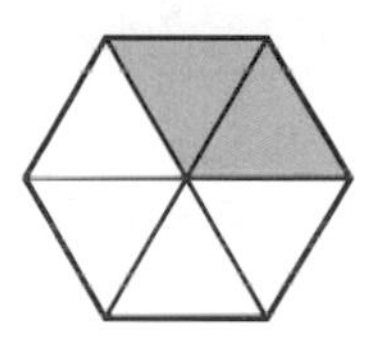

What fraction (in its simplest form) of the polygon is orange?

____

19 Anna is thinking of a number and gives the following clues:

"My number is:

- between 10 and 20
- a multiple of 3
- 1 less than a multiple of 4"

What number is Anna thinking of? ________

20 On this tower of bricks, the number on each brick is the sum of the numbers on the two bricks supporting it.

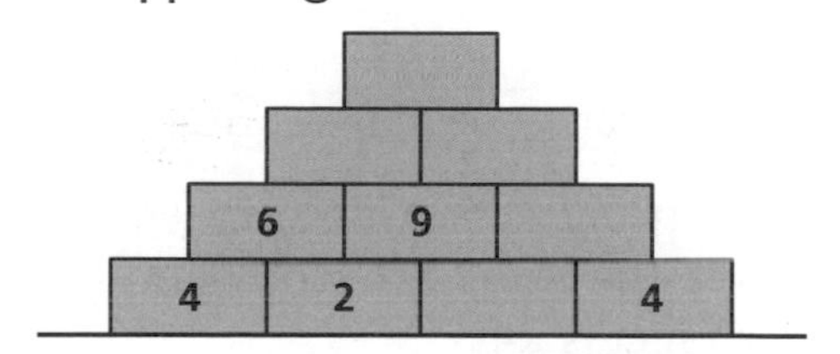

What number is on the top brick? ________

Test 10

Final score

20 ___ %

For all of the questions in this test, do the calculation entirely in your head with no written 'working' and just write down the answer.

1 Write in figures the number which is sixty-three more than twenty-seven. ___________

2 Cup-cakes are sold in packs of 6 and cost £1.45 per pack. Polly needs 18 cup-cakes for her classmates. What will be the total cost? £___________

3 Find the value of $3^2 - 1^2$ ___________

4 What is the area of a rectangle measuring 12 cm by 6 cm? ___________ cm²

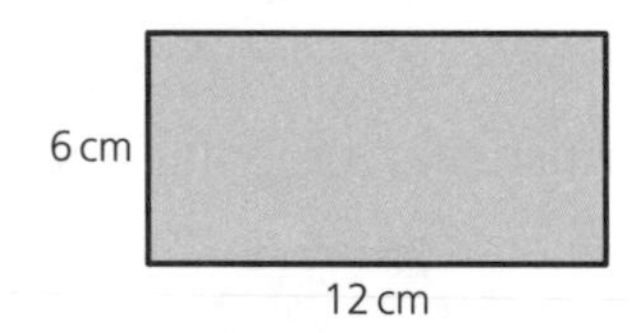

5 A packet of crisps contains 8 grams of fat. How many grams of fat are there in 4 packets of crisps? ___________ g

6 A DVD normally costs £10.00

The price is reduced by 10% in a sale. What is the sale price? £___________

7 Riley won a cross-country race with a time of 23 minutes and 10 seconds. Ed finished 1 minute and 8 seconds behind Riley. What was Ed's time?

_________ minutes _________ seconds

8 What is $3 \times 5 \times 8$? ___________

9 Isla buys 4 CDs costing £4.50 each. How much change does she receive from a £20 note?

£___________

10 What is a quarter of 360? ___________

11 Sara is thinking of two integers (whole numbers). The sum of the integers is 10 and their product is 24

What is the smaller number? ___________

12 Clarissa has the five number cards below.

What is the smallest odd number she can make by placing all five cards side by side?

____

13 The Carroll diagram shows some information about a group of people.

	Men	Women
Do not wear glasses	4	7
Wear glasses	8	5

How many people in the group wear glasses?

____

14 Tim achieves the following marks in a series of tests.

8 7 9 6 9 8 7 9 6 8 7 9 6

What mark is the mode? ___________

15 Which number between 60 and 70 is divisible exactly by 9? ___________

16 What is the smallest number of British coins needed to pay exactly 99 pence? ___________

17 Alberto has 48 marbles. 8 are blue and a quarter of the others are green. The rest are red. How many marbles are red? ___________

18 How many of the following calculations give a result with a units digit 8? ___________

$2 \times 39 \quad 56 \div 7 \quad 39 - 11 \quad 49 + 19$

19 Laura's birthday is 28 July. Sarah's is 2 August. Laura's birthday is on Sunday. On which day is Sarah's birthday? ___________

20 What is the sum of the first three square numbers?

____

Test 11

Final score

$\overline{20}$ ___ %

For all of the questions in this test, do the calculation entirely in your head with no written 'working' and just write down the answer.

In questions 1 to 10 you are reminded of 10 useful strategies which may help you in later questions.

1 7×21 ________

✱ Try 7×20 and then add 7

2 £$156 \div 4$ £________

✱ Divide by 2 and then by 2 again

3 49×5 ________

✱ Multiply by 10 then divide by 2

4 $56 - 39$ ________

✱ Subtract 40 then add 1

5 $156 \div 12$ ________

✱ Divide by 3, then by 4

6 1.5×4 ________

✱ You know that $15 \times 4 = 60$

7 $\frac{2}{5}$ of 40 ________

✱ Find $\frac{1}{5}$ then multiply by 2

8 $37 + 21 + 39$ ________

✱ $21 + 39 = 60$ first

9 2.9×4

✱ The result must be about 12

10 11×13 ________

✱ The units digit of the result is 3

11 Yuri has 25 sweets in his pocket. He eats 7 sweets and gives away 9

How many sweets does Yuri have left?

____

12 There are 14 people on a bus. 6 get on and 9 get off. How many people are now on the bus?

____

13 What is the cost of 7 chocolate bars priced at 48 pence each?

£________

14 Sally left home at 08:45 and reached the shop at 09:09. How long did the journey take her?

________ minutes

15 What is the largest number, less than 60, which divides exactly by 8?

____

16 Given that $374 \div 22 = 17$, what is $374 \div 11$?

____

17 By how much is 5×12 greater than 2×15?

____

18 The diagram shows a pattern of rectangles.

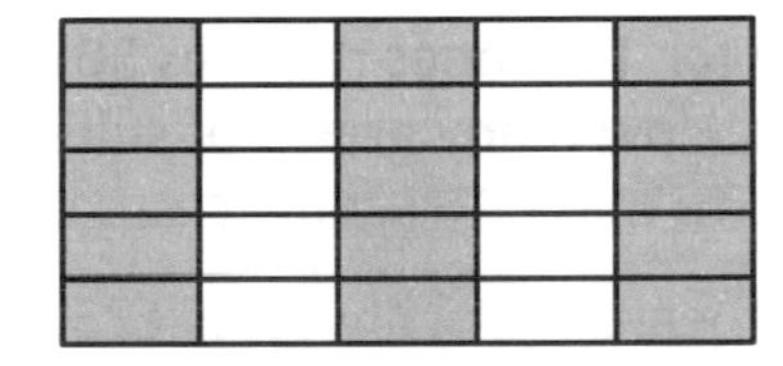

What percentage of the rectangles is blue?

________ %

19 Helena is thinking of a number and gives the following clues:

"My number is:

- between 40 and 50
- prime
- 1 less than a multiple of 7"

What number is Helena thinking of? ________

20 On this tower of bricks, the number on each brick is the sum of the numbers on the two bricks supporting it.

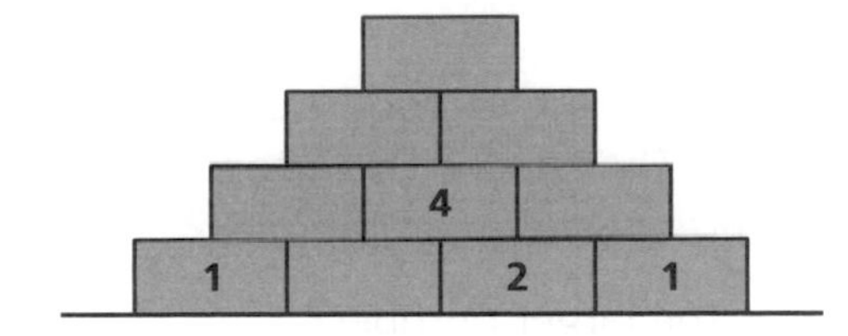

What number is on the top brick? ________

Test 14

Final score

/20 ___ %

For all of the questions in this test, do the calculation entirely in your head with no written 'working' and just write down the answer.

1 Write in figures the number which is twenty-seven less than sixty-two. ________

2 Mince pies are sold in packs of 12 and cost £2.30 per pack. Emma needs 48 mince pies for the Christmas party. What will be the total cost? £________

3 Find the value of $5^2 + 4^2$ ________

4 What is the perimeter of a rectangle measuring 23 cm by 8 cm? ________ cm

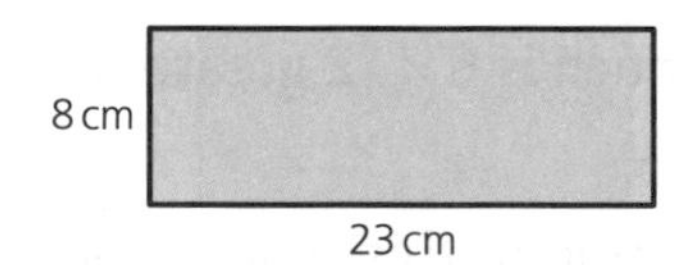

5 When Jo makes soup for 4 people she uses 800 ml of water. How many millilitres of water should she use to make soup for 3 people? ________ ml

6 A computer game normally costs £35.00

The price is reduced by 10% in a sale. What is the sale price? £________

7 Jon finished second in a cross-country race with a time of 28 minutes and 45 seconds. Connie finished 1 minute and 58 seconds ahead of Jon. What was Connie's time?

________ minutes ________ seconds

8 What is $5 \times 4 \times 4$? ________

9 June buys 2 DVDs costing £9.50 each. How much change does she receive from a £20 note?

£________

10 What is a third of 72? ________

11 Tyler is thinking of two integers (whole numbers), both less than 10. The product of the integers is 35

What is the larger number? ________

12 Victoria has the five number cards below.

What is the largest odd number she can make by placing all five cards side by side?

____

13 The Carroll diagram shows some information about a group of people.

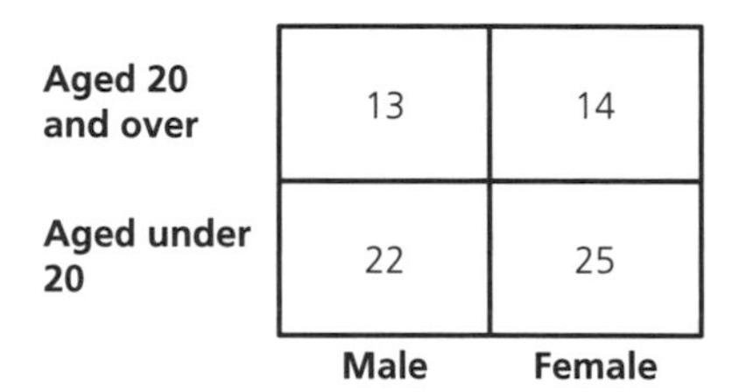

	Male	Female
Aged 20 and over	13	14
Aged under 20	22	25

How many people are in the group?

____

14 Dmitri's scores on 12 spelling tests are:

9 10 7 9 7 8 7 9 8 10 6 9

What score is the mode? ________

15 Which number between 50 and 70 is divisible exactly by 12? ________

16 What is the smallest number of British coins needed to pay exactly 55 pence? ________

17 Gabriel has 30 marbles. 6 are red and half of the others are blue. The rest are yellow. How many marbles are yellow?

____

18 How many of the following calculations give a result with a units digit 8?

____

4×32 $108 \div 9$ $56 - 22$ $56 + 102$

19 On 30 April, Gerda's age was recorded as 9 years 11 months. In which month is her birthday? ________

20 What is the sum of the numbers 20 to 22 inclusive?

____

Test 15

Final score

$\overline{}$ 20 ___ %

For all of the questions in this test, do the calculation entirely in your head with no written 'working' and just write down the answer.

In questions 1 to 10 you are reminded of 10 useful strategies which may help you in later questions.

1 19 × 21 ________

✱ Try 19 × 20 and then add 19

2 £30 ÷ 4 £________

✱ Divide by 2 and then by 2 again

3 63 × 5 ________

✱ Multiply by 10 then divide by 2

4 108 − 47 ________

✱ Subtract 50 then add 3

5 360 ÷ 15 ________

✱ Divide by 3, then by 5

6 0.7 × 6 ________

✱ You know that 7 × 6 = 42

7 $\frac{2}{3}$ of 240 ________

✱ Find $\frac{1}{3}$ then multiply by 2

8 38 + 9 + 12 ________

✱ 38 + 12 = 50 first

9 1.9 × 7 ________

✱ The result must be about 14

10 13 × 13 ________

✱ The units digit of the result is 9

11 Sergei has fourteen 50p coins in his pocket. He spends £3.50 and loses one 50p coin down a drain. How many coins does Sergei have left?

____

12 There are 56 people at a concert. 4 more people arrive late and 3 leave early. How many people are now at the concert?

____

13 What is the cost of 6 ice-creams priced at £1.20 each?

£________

14 Andy left home at 07:58 and reached the school at 08:45. How long did the journey take him?

________ minutes

15 What is the largest number, less than 80, which divides exactly by 3?

____

16 Given that 368 ÷ 23 = 16, what is 368 ÷ 16?

____

17 By how much is 7 × 15 greater than 5 × 17?

____

18 The diagram shows a pattern of rectangles.

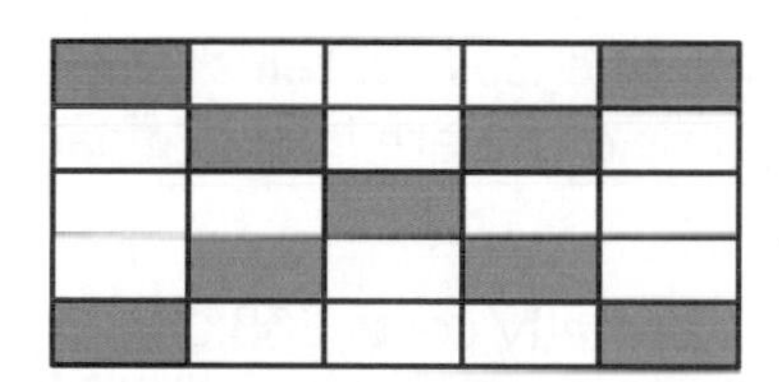

What percentage of the rectangles is white?

________ %

19 Heather is thinking of a number and gives the following clues:

"My number is:

- between 19 and 29
- 1 more than a prime number
- a multiple of 6"

What number is Heather thinking of? ________

20 On this tower of bricks, the number on each brick is the sum of the numbers on the two bricks supporting it.

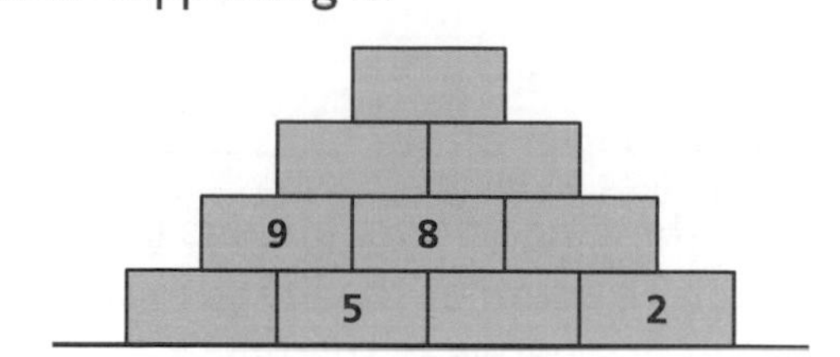

What number is on the top brick? ________

Test 16

Final score

$\overline{}$
20 ___ %

For all of the questions in this test, do the calculation entirely in your head with no written 'working' and just write down the answer.

1 Write in figures the number which is seventeen less than fifty-three. __________

2 Foreign stamps are sold in packs of fifty, priced at 90p per pack. Sandy wants to buy 1000 stamps. How much will the stamps cost? £__________

3 Find the value of $5^2 - 4^2$ __________

4 What is the area of a rectangle measuring 12 cm by 5 cm? __________ cm^2

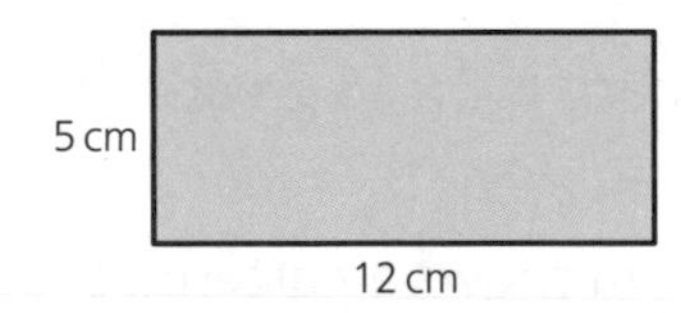

5 100 grams of crisps contains 32 g of fat. How many grams of fat are in a 25 g packet of crisps? __________ g

6 A jacket normally costs £68.00

The price is reduced by a 10% in a sale. What is the sale price?

£__________

7 Damien came second in a running race with a time of 3 minutes and 12 seconds. Cameron's winning time was 23 seconds faster. What was Cameron's time?

_______ minutes _______ seconds

8 What is $5 \times 6 \times 7$? __________

9 Mrs Grant buys 4 cups of tea costing 95p each. How much change does she receive from a £5 note?

£__________

10 What is a quarter of 56? __________

11 Cassie is thinking of two integers (whole numbers). The sum of the integers is 40 and the difference between them is 2

What is the larger number? __________

12 Hal has the five number cards below.

What is the smallest 4-digit odd number he can make by placing four of the cards side by side? __________

13 The Venn diagram shows some information about children in Year 5

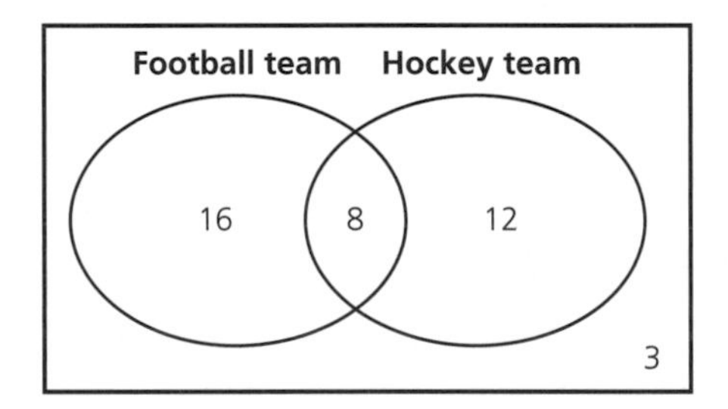

How many children in Year 5 have played in the football team but not in the hockey team? __________

14 What is the mode in the following set of scores? __________

7 5 9 6 7 8 9 4 7
5 9 6 7 9 8 7 5 6

15 Which number between 30 and 50 is divisible exactly by 13? __________

16 What is the smallest number of British coins needed to pay exactly 88 pence? __________

17 Jean has 40 sweets. A quarter are mints and a third of the others are toffees. The rest are chocolates. How many chocolates are there?

____

18 How many of the following calculations give a result with a units digit 4? __________

7×12 $48 \div 4$ $81 - 27$ $39 + 25$

19 Sam's birthday is 29 November and Tim's is 5 days earlier. Tim's birthday is on a Sunday. On which **day of the week** is Sam's birthday?

____

20 What is the sum of the numbers 18 to 20 inclusive? __________

Test 17

Final score

$\overline{20}$ ____ %

For all of the questions in this test, do the calculation entirely in your head with no written 'working' and just write down the answer.

In questions 1 to 10 you are reminded of 10 useful strategies which may help you in later questions.

1 7×48 ________

✱ Try 7×50 and then subtract 14

2 £20.60 × 4 £________

✱ Double and then double again

3 $210 \div 5$ ________

✱ Multiply by 2 then divide by 10

4 $47 + 79$ ________

✱ Add 80 then subtract 1

5 $432 \div 12$ ________

✱ Divide by 3, then by 4

6 1.3×4 ________

✱ You know that $13 \times 4 = 52$

7 $\frac{4}{5}$ of 50 ________

✱ Find $\frac{1}{5}$ then multiply by 4

8 $39 + 45 + 21$ ________

✱ $39 + 21 = 60$ first

9 29×4 ________

✱ The result must be about 120

10 13×5 ________

✱ The units digit of the result is 5

11 Shanna has 40 pairs of shoes. She gives 14 pairs to a charity shop and throws 3 pairs away. How many pairs of shoes does she have now?

____

12 Mike has 33 sweets. He gives a third of them to his brother and eats half of the remainder. How many sweets does he have left?

____

13 What is the cost of 8 CDs priced at £6.50 each?

£________

14 Dougal left home at 18:25 to walk to the theatre. The journey took 55 minutes. At what time did he reach the theatre?

________ : ________

15 What is the smallest number, greater than 100, which divides exactly by 6?

____

16 Given that $49 \times 8 = 392$, what is 49×4?

____

17 By how much is 5×14 greater than 4×15?

____

18 What percentage of this shape is purple?

________ %

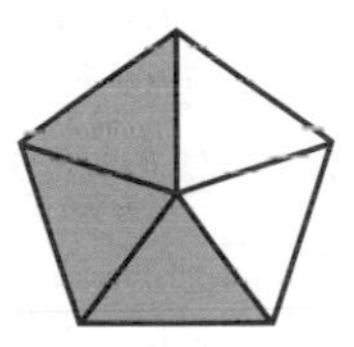

19 Sylvia is thinking of a number and gives the following clues:

"My number is:

- a multiple of 3
- a multiple of 7
- between 40 and 60"

What number is Sylvia thinking of? ________

20 On this tower of bricks, the number on each brick is the sum of the numbers on the two bricks supporting it.

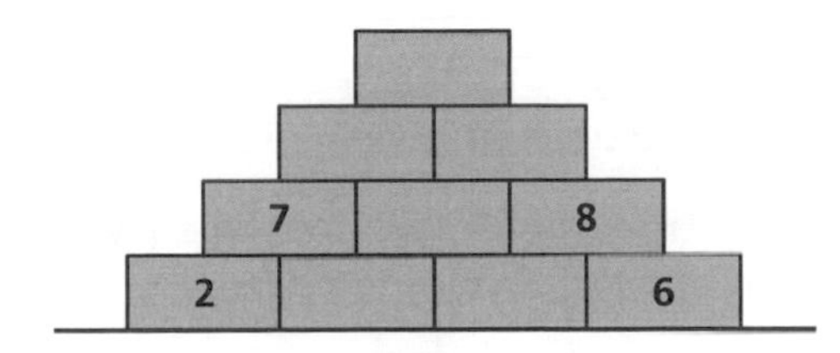

What number is on the top brick? ________

Test 18

Final score

$\overline{20}$ ___ %

For all of the questions in this test, do the calculation entirely in your head with no written 'working' and just write down the answer.

1 Write in figures the number which is thirty-seven more than forty-nine. ________

2 Scones are sold in packs of 4 and cost £1.40 per pack. Emma needs 20 scones for a coffee morning. What will be the total cost? £________

3 Find the value of $7^2 + 1^2$ ________

4 What is the perimeter of a rectangle measuring 15 cm by 5 cm? ________ cm

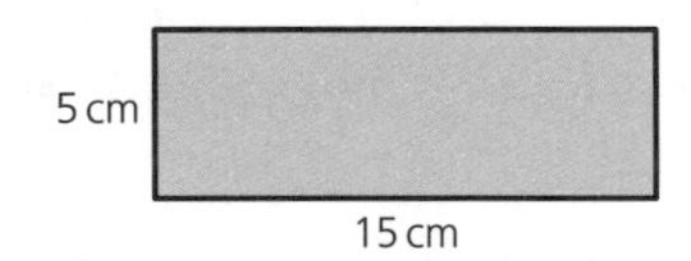

5 Sandra usually mixes 900 ml of water with 100 ml of concentrate to make a litre of orange juice. She has only 50 ml of concentrate. How much water should she use? ________ ml

6 A microwave oven normally costs £48.00

The price is reduced by 25% in a sale. What is the sale price? £________

7 Jenny finished first in a cross-country race with a time of 31 minutes and 31 seconds. May finished 47 seconds behind Jenny. What was May's time?

______ minutes ______ seconds

8 What is $8 \times 9 \times 10$? ________

9 Sharon buys 3 DVDs costing £10.95 each. How much change does she receive from a £50 note? £________

10 What is a quarter of 280? ________

11 Robbie is thinking of two integers (whole numbers) both less than 20. Their product is 55

What is the sum of the numbers? ________

12 Victoria has the four number cards below.

What is the smallest 3-digit even number she can make by placing three of the cards side by side? ________

13 The Carroll diagram shows some information about a group of people.

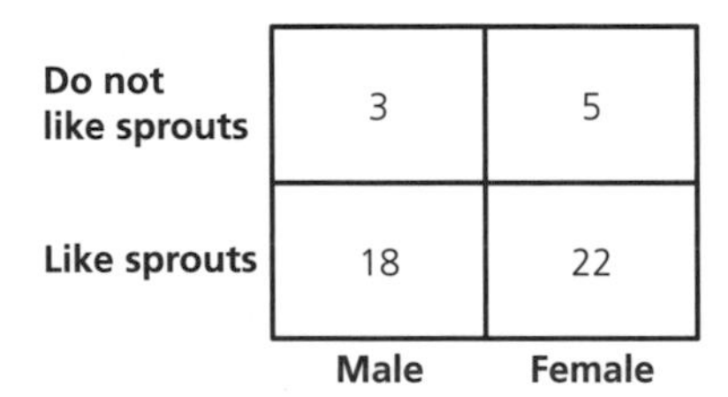

How many more people like sprouts than do not like sprouts? ________

14 What number is exactly half way between 20 and 32? ________

15 Which number between 40 and 60 is divisible exactly by 13? ________

16 What is the smallest number of British coins needed to pay exactly 19 pence?

____

17 Pria has 18 marbles. 6 are blue and a quarter of the others are red. The rest are green. How many marbles are green?

____

18 How many of the following calculations give a result with a units digit 0?

____

5×26 $60 \div 5$ $505 - 25$ $57 + 103$

19 On 31 December, Jill's age was recorded as 10 years 2 months. In which month is her birthday? ____________

20 What is the sum of the numbers 34 to 36 inclusive?

____

Test 19

Final score

$\overline{20}$ ___ %

For all of the questions in this test, do the calculation entirely in your head with no written 'working' and just write down the answer.

In questions 1 to 10 you are reminded of 10 useful strategies which may help you in later questions.

1 49×11 ________

* Try 49×10 and then add 49

2 £56 ÷ 4 £________

* Divide by 2 and then by 2 again

3 33×5 ________

* Multiply by 10 then divide by 2

4 $121 - 38$ ________

* Subtract 40 then add 2

5 $168 \div 12$ ________

* Divide by 2, then by 6

6 1.2×7 ________

* You know that $12 \times 7 = 84$

7 $\frac{3}{4}$ of 64 ________

* Find $\frac{1}{4}$ then multiply by 3

8 $73 + 19 + 27$ ________

* $73 + 27 = 100$ first

9 10.1×4 ________

* The result must be about 40

10 7×18 ________

* The units digit of the result is 6

11 Yuri had 16 sweets in his pocket. He ate half of them and gave a few to a friend. Yuri finds that he has 3 sweets left. How many did he give to his friend?

____

12 There are 7 people at a bus stop. 5 more people arrive but there are only 9 seats available on the bus. How many people remain at the bus stop?

____

13 What is the cost of 7 muffins priced at £1.05 pence each?

£________

14 Gloria left home at 17:45 and reached the Guide hall at 18:18. How long did the journey take her?

________ minutes

15 What is the largest number, less than 70, which divides exactly by 3? ________

16 Given that $80 \div 16 = 5$, what is $40 \div 16$?

____

17 By how much is 7×13 greater than 3×17?

____

18 The diagram shows a pattern of rectangles.

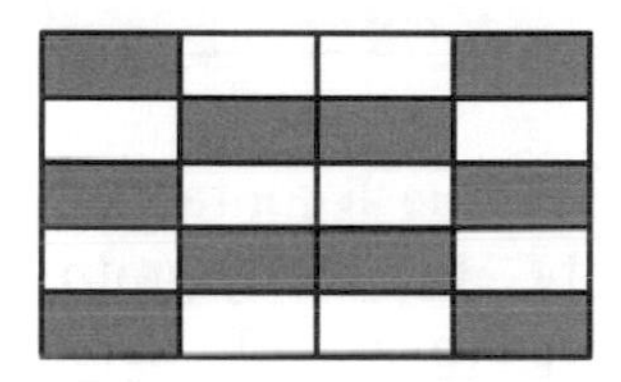

What percentage of the rectangles is orange?

________ %

19 Sean is thinking of a number and gives the following clues:

"My number is:

- between 50 and 70
- 1 more than a multiple of 7
- a multiple of 3"

What number is Sean thinking of? ________

20 On this tower of bricks, the number on each brick is the sum of the numbers on the two bricks supporting it.

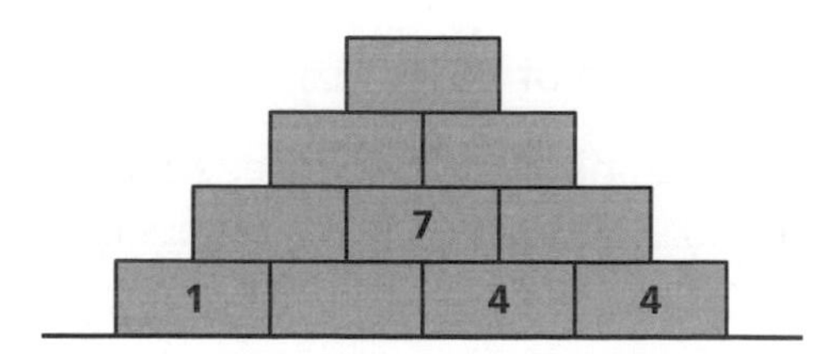

What number is on the top brick? ________

Test 20

Final score

$\overline{20}$ ___ %

For all of the questions in this test, do the calculation entirely in your head with no written 'working' and just write down the answer.

1 Write in figures the number which is seven less than forty-two. __________

2 Foreign coins are sold in bags of 20 and cost £1.20 per pack. Sandy wants to buy 100 coins. How much will the coins cost? £__________

3 Find the value of $6^2 - 4^2$ __________

4 What is the area of a rectangle measuring 11 cm by 7 cm? __________ cm²

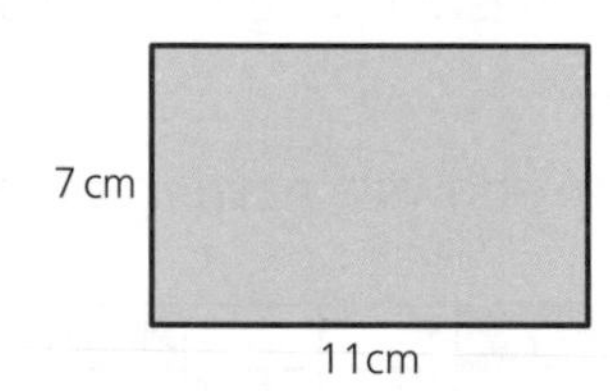

5 100 grams of muesli contains 60 g of carbohydrate. How many grams of carbohydrate are in a 40 g helping of muesli? __________ g

6 A sofa normally costs £700

The price is reduced by a 10% in a sale. What is the sale price? £__________

7 Yusuf came second in a swimming race with a time of 3 minutes and 3 seconds. Camille's winning time was 5 seconds faster. What was Camille's time?

________ minutes ________ seconds

8 What is $6 \times 7 \times 10$? __________

9 Mrs Taylor buys 4 bags of potatoes costing £2.40 each. How much change does she receive from a £20 note? £__________

10 What is a fifth of 85? __________

11 Connie is thinking of two integers (whole numbers). The sum of the integers is 12 and their product is 32

What is the larger number? __________

12 Maya has the five number cards below.

What is the largest 3-digit odd number she can make by placing three of the cards side by side? __________

13 The Venn diagram shows some information about children in Year 5

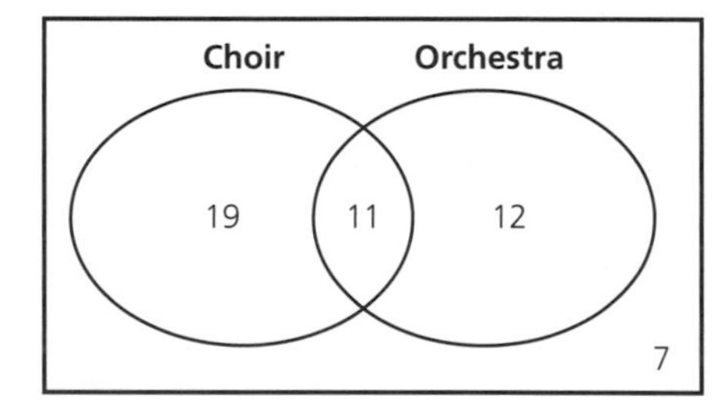

How many children in Year 5 do not sing in the choir? __________

14 What is the mode in the following set of scores? __________

18 19 14 11 15 19 16 17 18 17
15 15 19 16 17 19

15 Which number between 60 and 80 is divisible exactly by 14? __________

16 What is the smallest number of British coins needed to pay exactly 93 pence? __________

17 Marie has 45 sweets. A fifth are mints and a quarter of the others are toffees. The rest are chocolates. How many chocolates are there?

____

18 How many of the following calculations give a result with a units digit 9? __________

6×13 $54 \div 6$ $42 - 23$ $47 + 58$

19 Felix's birthday is 14 June and Tara's is 5 days later. Felix's birthday is on Tuesday. On which **day of the week** is Tara's birthday?

____

20 What is the sum of the numbers 25 to 27 inclusive? __________

Test 21

Final score

$\overline{20}$ ___ %

For all of the questions in this test, do the calculation entirely in your head with no written 'working' and just write down the answer.

In questions 1 to 10 you are reminded of 10 useful strategies which may help you in later questions.

1 9×49 ________

* Try 9×50 and then subtract 9

2 $26\,\text{kg} \times 4$ ________ kg

* Double and then double again

3 $160 \div 5$ ________

* Multiply by 2 then divide by 10

4 $53 + 68$ ________

* Add 70 then subtract 2

5 $300 \div 15$ ________

* Divide by 3, then by 5

6 1.2×8 ________

* You know that $12 \times 8 = 96$

7 $\frac{3}{4}$ of 120 ________

* Find $\frac{1}{4}$ then multiply by 3

8 $44 + 79 + 36$ ________

* $44 + 36 = 80$ first

9 19×5

* The result must be about 100

10 13×8 ________

* The units digit of the result is 4

11 Dev has a case of 40 bottles of lemonade. He sells 10% of them and gives 8 bottles to his children. How many bottles does he have left?

____

12 Simon has 30 marbles. He loses 7 to Ben and wins 13 from Jenny. How many marbles does he have now?

____

13 What is the cost of 6 DVDs priced at £11.50 each?

£________

14 Omar left home at 07:55 to drive to work. The journey took 37 minutes. At what time did he reach work?

________ : ________

15 What is the smallest number, greater than 50, which is a multiple of both 3 and 4?

____

16 Given that $56 \times 6 = 336$, what is 56×2?

____

17 By how much is 6×13 greater than 3×16?

____

18 What fraction of this shape is blue? ________

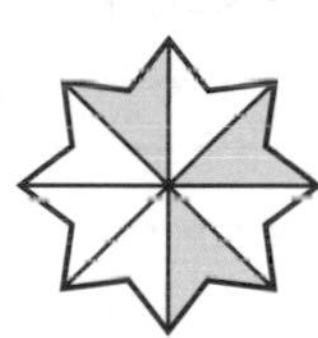

19 Serena is thinking of a number and gives the following clues:

"My number is:

- even
- a multiple of 9
- between 40 and 60"

What number is Serena thinking of? ________

20 On this tower of bricks, the number on each brick is the sum of the numbers on the two bricks supporting it.

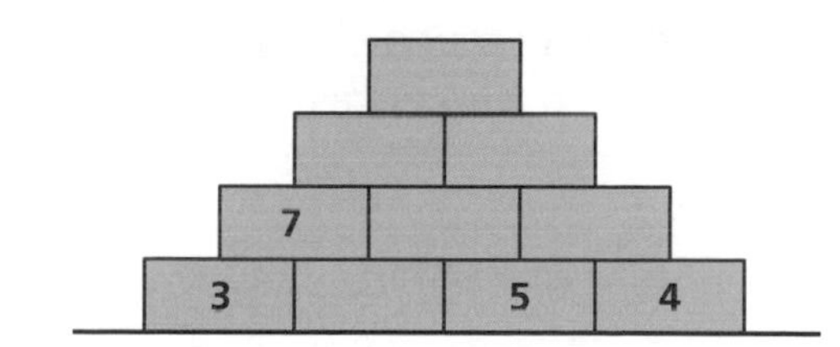

What number is on the top brick? ________

Test 22

Final score

$\overline{20}$ ___ %

For all of the questions in this test, do the calculation entirely in your head with no written 'working' and just write down the answer.

1 Write in figures the number which is fifty-eight less than eighty-five. ________

2 Cup-cakes are sold in packs of 6 and cost £2.10 per pack. Annie needs 30 cup-cakes for a party. What will be the total cost? £________

3 Find the value of $10^2 + 2^2$ ________

4 What is the perimeter of a rectangle measuring 14 cm by 6 cm? ________ cm

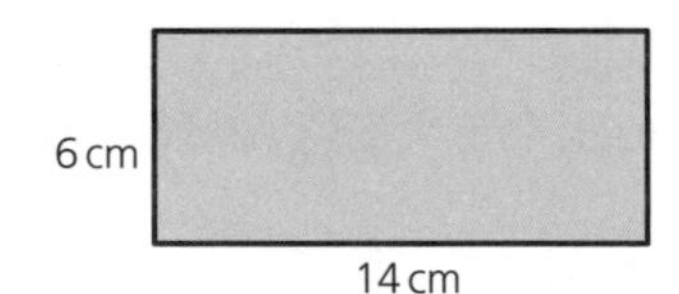

5 Mel usually mixes 840 ml of water with 160 ml of concentrate to make a litre of lemon juice. She has only 80 ml of concentrate. How much water should she use? ________ ml

6 A toaster normally costs £16.80

The price is reduced by 10%. What is the new price? £________

7 Amy finished first in a cross-country ski race with a time of 45 minutes and 11 seconds. Tim finished 2 minutes and 17 seconds behind Amy. What was Tim's time?

________ minutes ________ seconds

8 What is $7 \times 4 \times 5$? ________

9 Celine buys 2 bottles of perfume costing £35.50 each. How much change does she receive from two £50 notes?

£________

10 What is a third of 450? ________

11 Tammy is thinking of two integers (whole numbers). Their product is 37

What is the sum of the numbers? ________

12 Mike has the five number cards below.

What is the largest 4-digit odd number he can make by placing four of the cards side by side? ________

13 The Carroll diagram shows some information about a group of people.

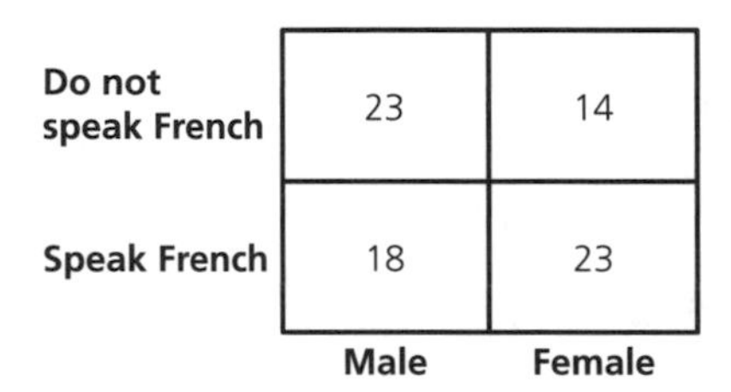

	Male	Female
Do not speak French	23	14
Speak French	18	23

How many more people in the group speak French than do not speak French? ________

14 What number is exactly half way between 44 and 54? ________

15 Which number between 50 and 60 is divisible exactly by 8? ________

16 What is the smallest number of British coins needed to pay exactly 43 pence?

____

17 Tina has 24 marbles. A third are red and half of the others are green. The rest are yellow. How many marbles are yellow?

____

18 How many of the following calculations give a result with a units digit 1?

____

9×19 $99 \div 9$ $98 - 37$ $54 + 107$

19 On 31 December, Jon's age was recorded as 10 years 5 months. In which month is his birthday? ________

20 What is the sum of the numbers 18 to 20 inclusive?

____

Test 23

Final score

$\overline{20}$ ___ %

For all of the questions in this test, do the calculation entirely in your head with no written 'working' and just write down the answer.

In questions 1 to 10 you are reminded of 10 useful strategies which may help you in later questions.

1 51×9 ________

* Try 50×9 and then add 9

2 £$120 \div 8$ £________

* Divide by 2, by 2 again and then by 2 again

3 47×5 ________

* Multiply by 10 then divide by 2

4 $96 - 49$ ________

* Subtract 50 then add 1

5 $154 \div 14$ ________

* Divide by 2, then by 7

6 0.4×9 ________

* You know that $4 \times 9 = 36$

7 $\frac{2}{5}$ of 45 ________

* Find $\frac{1}{5}$ then multiply by 2

8 $57 + 33 + 26$ ________

* $57 + 33 = 90$ first

9 19.9×2 ________

* The result must be about 40

10 5×24 ________

* The units digit of the result is 0

11 Andrei had 12 conkers. He gave 3 away and 4 shattered whilst he was playing conkers. How many conkers does he have left?

____

12 There are 15 children in a playground. 7 leave and a group arrives in a minibus. Now there are 22 children in the playground. How many arrived in the minibus?

____

13 What is the cost of 5 cup-cakes priced at 85 pence each?

£________

14 Alex left home at 11:53 and reached the shop at 12:17. How long did the journey take him?

________ minutes

15 What is the largest number, less than 100, which divides exactly by 4? ________

16 Given that $128 \div 16 = 8$, what is $128 \div 32$?

____

17 By how much is 8×12 greater than 2×18?

____

18 The diagram shows a pattern of tiles.

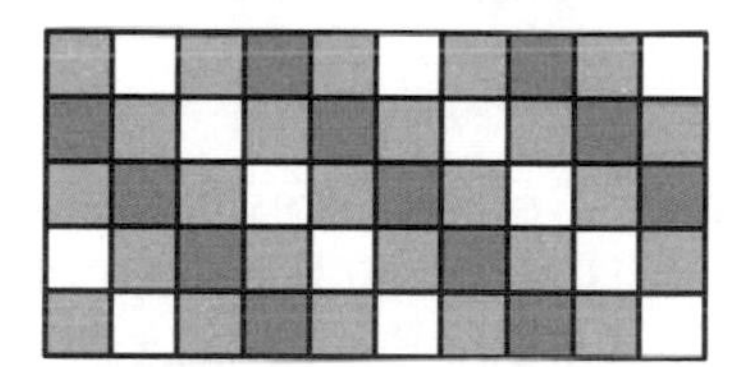

What percentage of the tiles is pink?

________ %

19 Rory is thinking of a number and gives the following clues:

"My number is:

- between 30 and 50
- 1 less than a square number
- a multiple of 5"

What number is Rory thinking of? ________

20 On this tower of bricks, the number on each brick is the sum of the numbers on the two bricks supporting it.

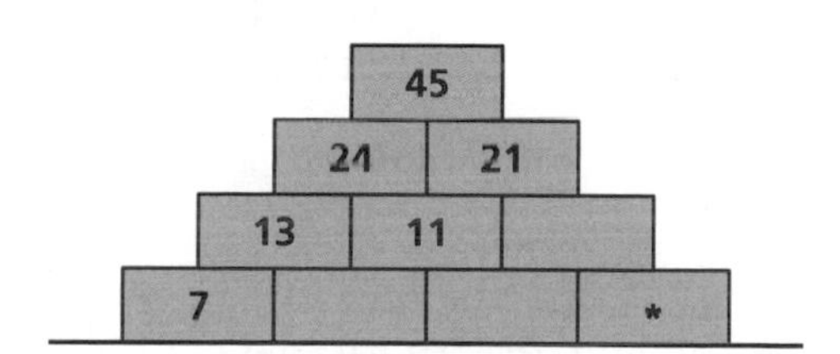

What number is on the brick marked with a star (*)? ________

Test 24

Final score

$\overline{20}$ ___ %

For all of the questions in this test, do the calculation entirely in your head with no written 'working' and just write down the answer.

1 Write in figures the number which is six less than one hundred and three. ________

2 Foreign stamps are sold in packets of 50 and cost 75p per packet. Keira wants to buy 1000 stamps. How much will the stamps cost? £________

3 Find the value of $3^2 - 2^2$ ________

4 What is the area of a rectangle measuring 13 cm by 3 cm? ________ cm²

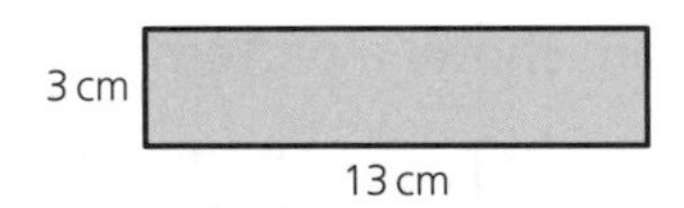

5 A 1 kg box of *Doggybickies* contains 90 biscuits. How many biscuits would there be in a 1.5 kg box of *Doggybickies*?

____

6 A freezer normally costs £450

The price is reduced by 20% in a sale. What is the sale price? £________

7 Diana came first in a cycle race with a time of 5 minutes and 44 seconds. Leanne came second with a time of 6 minutes and 3 seconds. By how many seconds did Diana beat Leanne?

________ seconds

8 What is $5 \times 12 \times 11$? ________

9 Mr Macdonald buys 5 bags of peat costing £4.80 each. How much change does he receive from two £20 notes?

£________

10 What is a sixth of 78? ________

11 Julie is thinking of two integers (whole numbers). The sum of the integers is 16 and the difference between them is 4

What is the larger number? ________

12 Romeo has the four number cards below.

What is the difference between the largest and the smallest 2-digit numbers he can make by placing two cards side by side?

____

13 The Venn diagram shows some information about children in Year 5

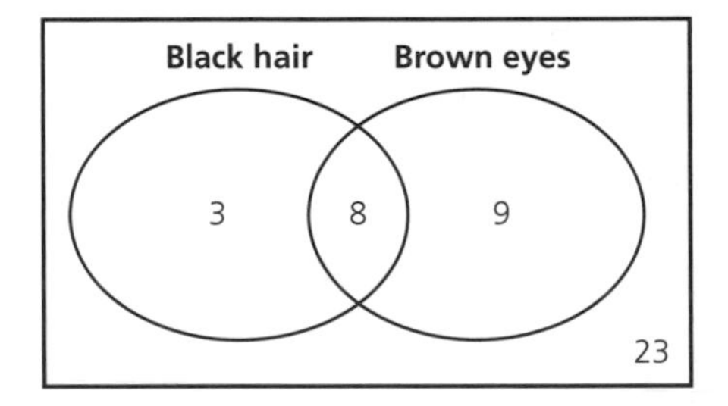

How many children in Year 5 do not have brown eyes? ________

14 What is the mode in the following set of scores? ________

9 6 7 9 4 8 7 6 4 9 3 8 8 9 4
1 5 9 6 7 8 7 5 5

15 Which number between 50 and 60 is divisible exactly by 7? ________

16 What is the smallest number of British coins needed to pay exactly £1.46? ________

17 Fabienne has a box of 24 chocolates. A third are dark chocolate and half of the others are milk chocolate. What fraction of the total number are milk chocolate? ________

18 How many of the following calculations give a result with a units digit 7? ________

7×17 $49 \div 7$ $22 - 5$ $35 + 124$

19 Lara's birthday is 3 January and India's is 6 days earlier. What date is India's birthday?

____

20 What is the sum of the numbers 50 to 52 inclusive? ________

Test 25

Final score

$\overline{20}$ ___ %

For all of the questions in this test, do the calculation entirely in your head with no written 'working' and just write down the answer.

In questions 1 to 10 you are reminded of 10 useful strategies which may help you in later questions.

1 19×9 ________

✱ Take easier steps when you multiply

2 $21\text{ cm} \times 8$ ________ cm

✱ Double or halve

3 $145 \div 5$ ________

✱ Use your 10s and 2s

4 $78 + 59$ ________

✱ Take easier steps when you add

5 $294 \div 14$ ________

✱ Use factors to divide

6 0.7×8 ________

✱ Use known facts

7 $\frac{4}{5}$ of 60 ________

✱ Use a step-by-step approach

8 $53 + 48 + 27$ ________

✱ Group when you add

9 29×4 ________

✱ Approximate the result

10 27×4 ________

✱ Check using the units digit of the result

11 Colin has a box of 24 chocolates. He eats a third of them and gives 6 to Mary. How many chocolates does he have left?

____

12 Damien has £3.60 in his pocket. He spends £1.95 and his aunt gives him a £2 coin. How much does he have now?

£________

13 What is the cost of 4 books priced at £5.95 each?

£________

14 Violet left home at 07:35 to do her paper round and got back at 08:29. How long was she away from home?

________ minutes

15 What is the smallest number, greater than 60, which is a multiple of both 3 and 5?

____

16 Given that $84 \times 5 = 420$, what is 168×5?

____

17 By how much is 7×12 greater than 2×17?

____

18 What fraction of this shape is green?

____

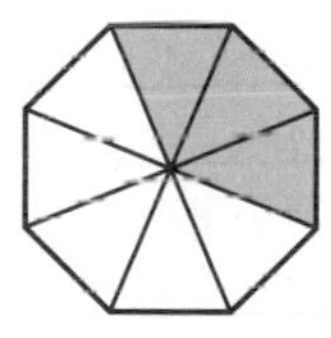

19 Clara is thinking of a number and gives the following clues:

"My number is:

- odd
- less than 20
- 1 more than a cube number."

What number is Clara thinking of? ________

20 On this tower of bricks, the number on each brick is the sum of the numbers on the two bricks supporting it.

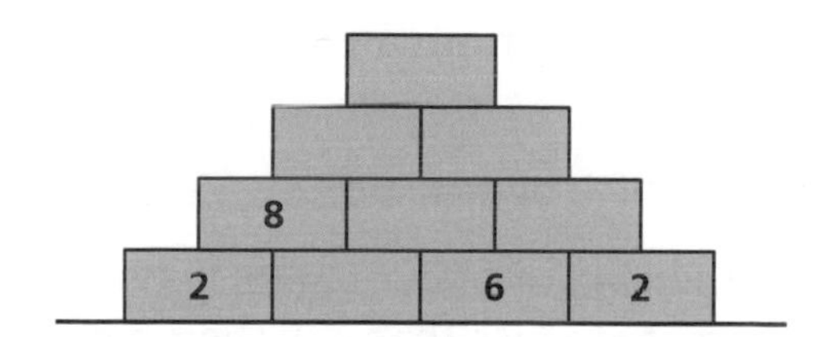

What number is on the top brick? ________

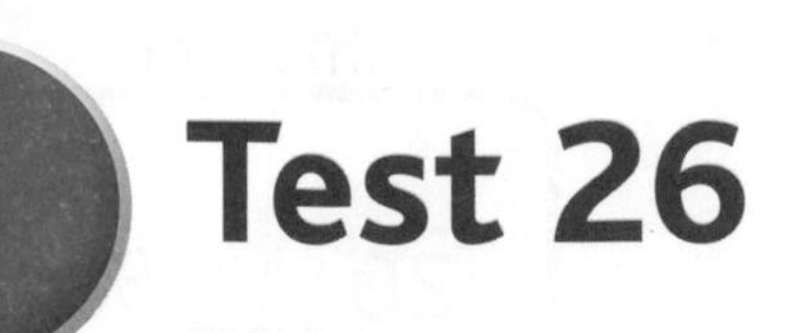

Test 26

Final score

$\overline{20}$ ___ %

For all of the questions in this test, do the calculation entirely in your head with no written 'working' and just write down the answer.

1 Write in figures the number which is sixty-one less than ninety-five. ___________

2 Rock cakes are sold in packs of 4 and cost £1.98 per pack. Juliet buys 12 rock cakes to share with her friends. What will be the total cost? £___________

3 Find the value of $11^2 + 1^2$ ___________

4 What is the perimeter of a rectangle measuring 13 cm by 8 cm? ___________ cm

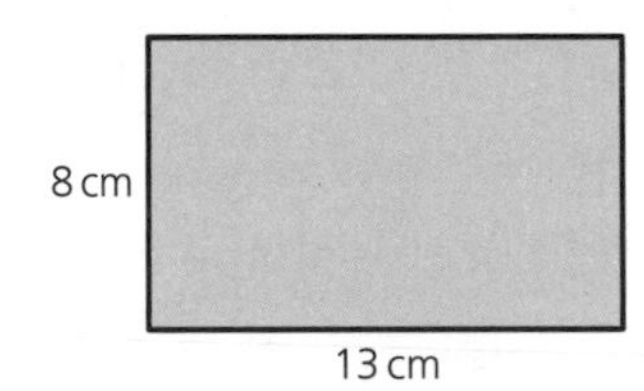

5 Theo mixes 400 ml of water with 100 ml of concentrate to make half a litre of orange juice to serve 2 people. How much concentrate will he need to make orange juice for 5 people? ___________ ml

6 A fridge normally costs £180

The price is reduced by 10%. What is the new price? £___________

7 Ari finished first in a sack race with a time of 45.8 seconds. Candice finished 1.7 seconds behind Ari. What was Candice's time?

_______ seconds

8 What is $2 \times 5 \times 35$? ___________

9 Miranda buys 2 charms costing £23.50 each for her charm bracelet. How much change does she receive from a £50 note? £___________

10 What is a fifth of 1 million? ___________

11 Ava is thinking of two integers (whole numbers) both less than 20. Their product is 34. What is the sum of the numbers?

____

12 Mandy has the four number cards below.

What is the largest 3-digit multiple of 5 she can make by placing three of the cards side by side? ___________

13 The Carroll diagram shows some information about a group of people.

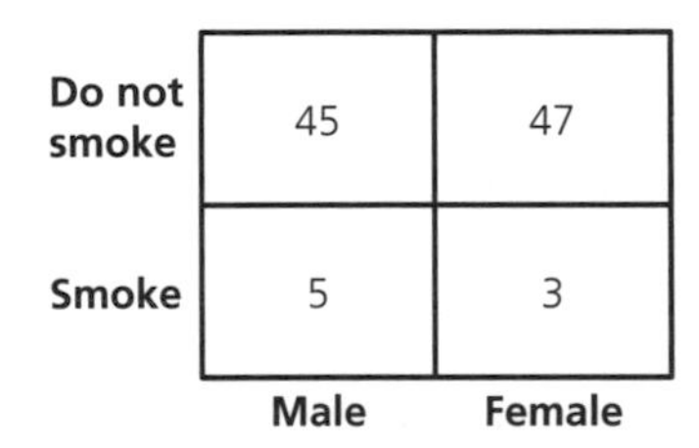

	Male	Female
Do not smoke	45	47
Smoke	5	3

What percentage of the people do not smoke?

___________ %

14 What number is exactly half way between 56 and 66? ___________

15 Which number between 70 and 79 is divisible exactly by 8? ___________

16 What is the smallest number of British coins needed to pay exactly £1.35? ___________

17 Trisha has 20 sweets. A quarter are toffees and 8 are mints. The rest are chocolates. How many chocolates are there? ___________

18 How many of the following calculations give a result with a units digit 8?

____

14×12 $96 \div 8$ $82 - 26$ $12 + 1006$

19 On 30 April, Ed's age was recorded as 11 years 3 months. In which month is his birthday?

____

20 What is the sum of the numbers 31 to 33 inclusive?

____

Test 27

Final score

$\overline{}$ / 20 ____ %

For all of the questions in this test, do the calculation entirely in your head with no written 'working' and just write down the answer.

In questions 1 to 10 you are reminded of 10 useful strategies which may help you in later questions.

1 69×6 ________

* Try 70×6 and then subtract 6

2 $240 \div 8$ ________

* Divide by 2, by 2 again and then by 2 again

3 53×5 ________

* Multiply by 10 then divide by 2

4 $100 - 57$ ________

* Subtract 60 then add 3

5 $165 \div 15$ ________

* Divide by 3, then by 5

6 1.2×7 ________

* You know that $12 \times 7 = 84$

7 $\frac{2}{3}$ of 66 ________

* Find $\frac{1}{3}$ then multiply by 2

8 $103 + 55 + 17$ ________

* $103 + 17 = 120$ first

9 99×3 ________

* The result must be about 300

10 7×14 ________

* The units digit of the result is 8

11 April had 12 beetles. 3 escaped and she caught 7 more. How many beetles does she have now? ________

12 40 friends were at a party. 60% of them were girls. How many boys were there?

____

13 What is the cost of a dozen muffins priced at £1.80 for four?

£________

14 Amanda left school at 15:55 and reached home at 16:35. How long did the journey take her?

________ minutes

15 What is the largest number, less than 100, which divides exactly by 9?

____

16 Given that $312 \div 8 = 39$, what is $312 \div 4$?

____

17 By how much is 9×18 greater than 8×19?

____

18 The diagram shows a pattern of tiles.

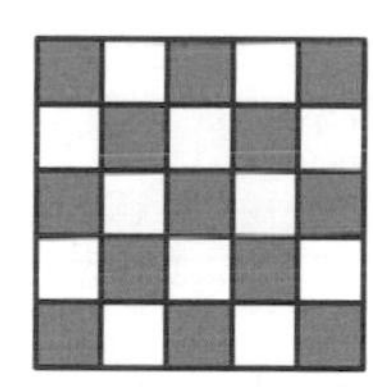

What percentage of the tiles is red?

________ %

19 Amber is thinking of a number and gives the following clues:

"My number is:

- between 40 and 50
- 1 more than a multiple of 6
- prime."

What number is Amber thinking of? ________

20 On this tower of bricks, the number on each brick is the sum of the numbers on the two bricks supporting it.

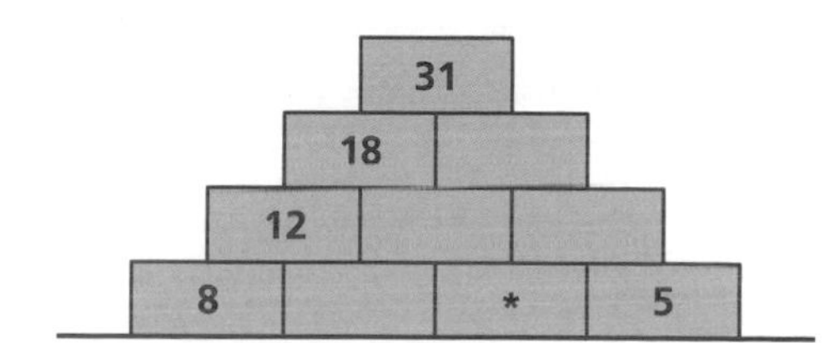

What number is on the brick marked with a star (*)?

____

Test 28

Final score

$\overline{}$
20 ____ %

For all of the questions in this test, do the calculation entirely in your head with no written 'working' and just write down the answer.

1 Write in figures the number which is eight less than two hundred and twelve.

____

2 Game cards are sold in packets of 10 and cost £1.45 per packet. Kevin wants to buy 200 cards. How much will the cards cost?
£__________

3 Find the value of $5^2 - 4^2$ __________

4 What is the area of a rectangle measuring 10 cm by 2.5 cm? __________ cm^2

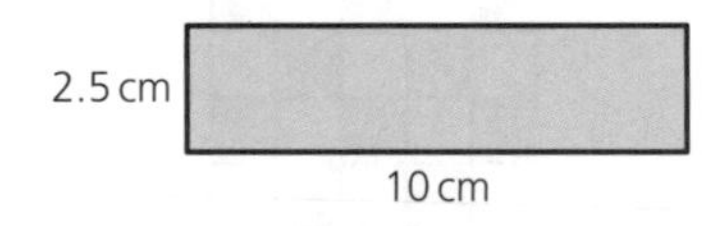

5 A 100 g packet of chocolate drops contains 120 chocolate drops. How many chocolate drops would there be in a 1 kg box of chocolate drops? __________

6 Mr Smith buys a car which normally costs £7500, but the price is reduced by 10%.

What does Mr Smith pay? £__________

7 Clare came second in a hopping race with a time of 2 minutes and 3 seconds. Lucy came first with a time of 1 minute and 49 seconds. By how many seconds did Lucy beat Clare?

________ seconds

8 What is $8 \times 18 \times 5$? __________

9 Dr Foster buys 3 bags of sticks costing £3.90 per bag. How much change does he receive from a £20 note?

£__________

10 What is a half of 17? __________

11 Jassica is thinking of two integers (whole numbers). The sum of the integers is 25 and the difference between them is 3

What is the smaller number? __________

12 Lucas has the four number cards below.

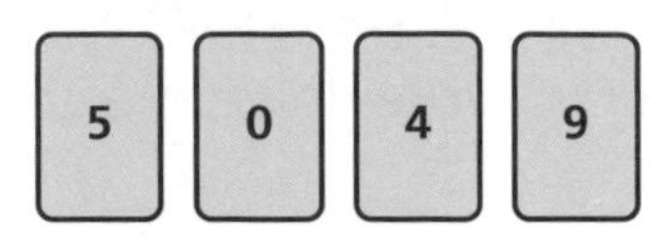

What is the largest multiple of 5 he can make by placing all four cards side by side?

____

13 The Venn diagram shows some information about children in a sports club.

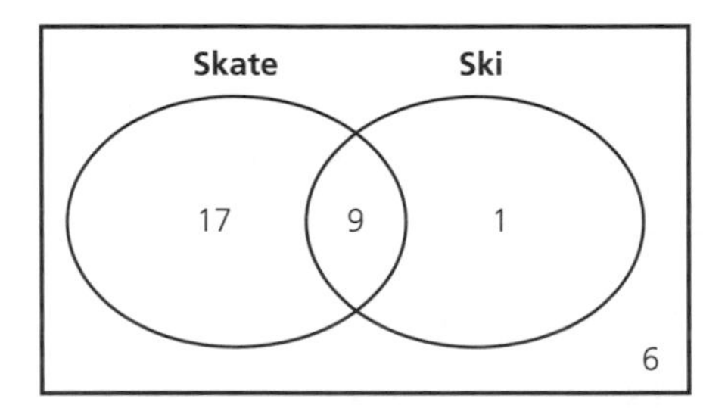

How many children in the club can ski or skate or do both sports? __________

14 What is the mode in the following set of scores? __________

9 6 9 4 8 7 7 8 9 5 9 6 4 7

15 Which number between 60 and 70 is divisible exactly by 8? __________

16 What is the smallest number of British coins needed to pay exactly £2.56? __________

17 Maria has a collection of dolls. She gives $\frac{1}{4}$ of them to her little sister and the rest to charity. The charity receives 9 dolls. How many dolls did Maria have? __________

18 How many of the following calculations give a result with a units digit 2?

____

6×14 $104 \div 2$ $53 - 35$ $49 + 73$

19 Zara's birthday is 27 November and Paul's is 5 days later. What date is Paul's birthday?

____

20 What is the sum of the numbers 39 to 41 inclusive? __________

Test 29

Final score

$\overline{20}$ ___ %

For all of the questions in this test, do the calculation entirely in your head with no written 'working' and just write down the answer.

In questions 1 to 10 you are reminded of 10 useful strategies which may help you in later questions.

1 31×11 __________

✱ Take easier steps when you multiply

2 33 cm × 4 __________ cm

✱ Double or halve

3 $130 \div 5$ __________

✱ Use your 10s and 2s

4 $148 + 39$ __________

✱ Take easier steps when you add

5 $147 \div 21$ __________

✱ Use factors to divide

6 0.9×12 __________

✱ Use known facts

7 $\frac{3}{5}$ of 80 __________

✱ Use a step-by-step approach

8 $75 + 16 + 25$ __________

✱ Group when you add

9 49×5 __________

✱ Approximate the result

10 33×6 __________

✱ Check using the units digit of the result

11 Cora has a bag of 18 sweets which she shares with Milly and Ben. Milly eats 5 sweets, Ben eats 7 sweets and Cora eats the rest. How many sweets does Cora eat?

____

12 Olivia has £5.73 in her purse. She spends £2.49 and finds a 50p coin in her pocket. How much does she have now?

£__________

13 What is the cost of 5 chocolate bars priced at 72p each?

£__________

14 Emily left home at 18:45 to attend a judo grading and got back at 20:25. How long was she away from home?

__________ hour(s) __________ minutes

15 What is the smallest number, greater than 100, which is a multiple of 8?

____

16 Given that $61 \times 22 = 1342$, what is 60×22?

____

17 By how much is 6×15 greater than 5×16?

____

18 What percentage of this shape is purple?

__________ %

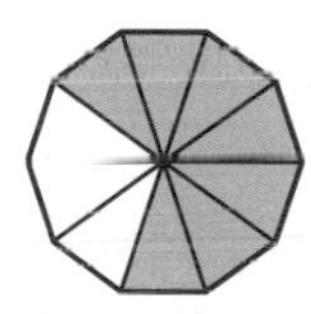

19 Amir is thinking of a 2-digit number and gives the following clues:

"My number is:

- even
- a multiple of 7

The units digit is 1 more than the tens digit."

What number is Amir thinking of? __________

20 On this tower of bricks, the number on each brick is the sum of the numbers on the two bricks supporting it.

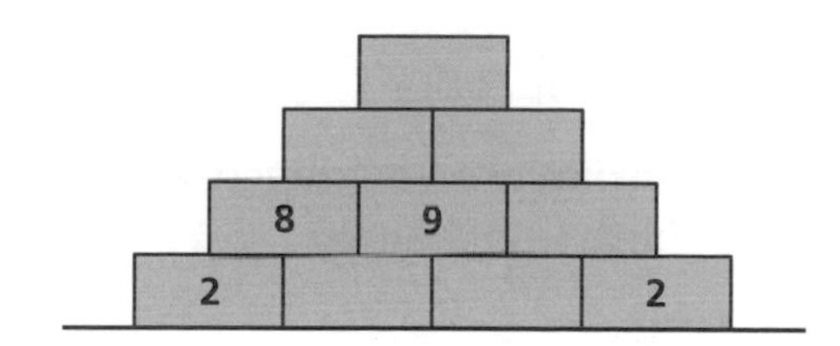

What number is on the top brick? __________

Test 30

Final score

$\overline{}$
20 ____ %

For all of the questions in this test, do the calculation entirely in your head with no written 'working' and just write down the answer.

1 Write in figures the number which is fifty-nine more than one hundred and five.

____

2 Small apple pies are sold in packs of 4 and cost £1.85 per pack. Christina buys 8 apple pies for a family picnic. What will be the total cost? £__________

3 Find the value of $10^2 + 5^2$ __________

4 What is the perimeter of a rectangle measuring 18 cm by 5 cm? __________ cm

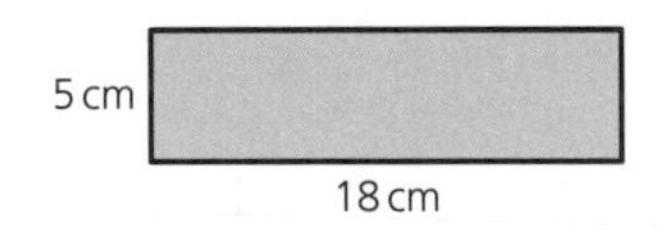

5 A pudding for 6 people requires 450 ml of milk. Lois wants to make the same pudding for 4 people. How much milk should she use?

__________ ml

6 A sweater normally costs £35

The price is reduced by 20%. What is the new price? £__________

7 Connie finished second in a swimming race with a time of 1 minute 4 seconds. Danni beat her by 7 seconds. What was Danni's time?

________ seconds

8 What is $4 \times 9 \times 5$? __________

9 Penny buys 2 pendants costing £9.75 each. How much change does she receive from a £20 note?

£__________

10 What is a fifth of 2000? __________

11 Maria is thinking of two integers (whole numbers) both less than 20

Their product is 35

What is the sum of the numbers? __________

12 Helen has the six number cards below.

3 | 4 | 5 | 6 | 4 | 5

What is the largest even number she can make by placing three of the cards side by side?

____

13 The Carroll diagram shows some information about animals in a small zoo.

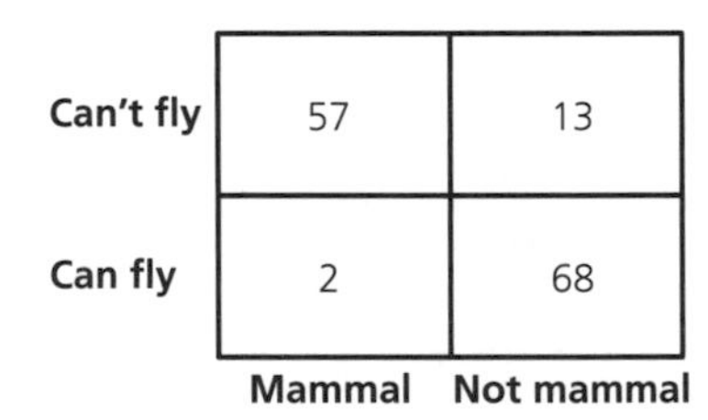

What fraction (in its simplest form) of the animals can fly?

____

14 What number is exactly half way between 30 and 48?

____

15 Which number between 70 and 80 is divisible exactly by 9?

____

16 What is the smallest number of British coins needed to pay exactly £2.50?

____

17 A garage has 40 cars. A quarter are red and 15 are blue. The rest are silver. How many cars are silver? __________

18 How many of the following calculations give a result with a units digit 4? __________

$13 \times 16 \quad 60 \div 15 \quad 32 - 16 \quad 45 + 29$

19 On 30 September, Eli's age was recorded as 10 years 9 months. In which month is his birthday? ______________

20 What is the sum of the numbers 35 to 37 inclusive? __________

Test 31

Final score

$\overline{}$ 20 ___ %

For all of the questions in this test, do the calculation entirely in your head with no written 'working' and just write down the answer.

In questions 1 to 10 you are reminded of 10 useful strategies which may help you in later questions.

1 89×4 ________

✱ *Try 90×4 and then subtract 4*

2 $320 \div 8$ ________

✱ *Divide by 2, by 2 again and then by 2 again*

3 63×5 ________

✱ *Multiply by 10 then divide by 2*

4 $200 - 169$ ________

✱ *Subtract 170 then add 1*

5 $156 \div 12$ ________

✱ *Divide by 3, then by 4*

6 0.9×9 ________

✱ *You know that $9 \times 9 = 81$*

7 $\frac{4}{5}$ of 70 ________

✱ *Find $\frac{1}{5}$ then multiply by 4*

8 $159 + 76 + 41$ ________

✱ *$159 + 41 = 200$ first*

9 199×4 ________

✱ *The result must be about 800*

10 9×23 ________

✱ *The units digit of the result is 7*

11 May had 20 woodlice in a choice chamber. 13 were in the dark side and the rest were in the light side. How many woodlice were in the light side?

____

12 There were 48 singers in the choir. $\frac{3}{4}$ of them were girls. How many boys were there?

____

13 What is the cost of 4 yoghurts priced at £1.65 for 2?

£________

14 Amelia left home at 07:50 and reached school at 08:25. How long did the journey take her?

________ minutes

15 What is the smallest number, greater than 50, which divides exactly by 8?

____

16 Given that $928 \div 29 = 32$, what is $928 \div 16$?

____

17 By how much is 8×13 greater than 3×18?

____

18 The diagram shows a pattern of tiles.

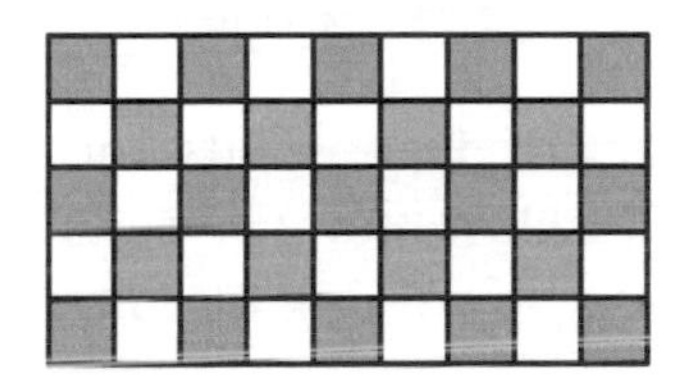

How many more white tiles are there than purple tiles?

________ more

19 Tamsin is thinking of a number and gives the following clues:

"My number is:

- less than 30
- 1 more than a multiple of 5
- prime."

What number is Tamsin thinking of? ________

20 On this tower of bricks, the number on each brick is the **product** of the numbers on the two bricks supporting it.

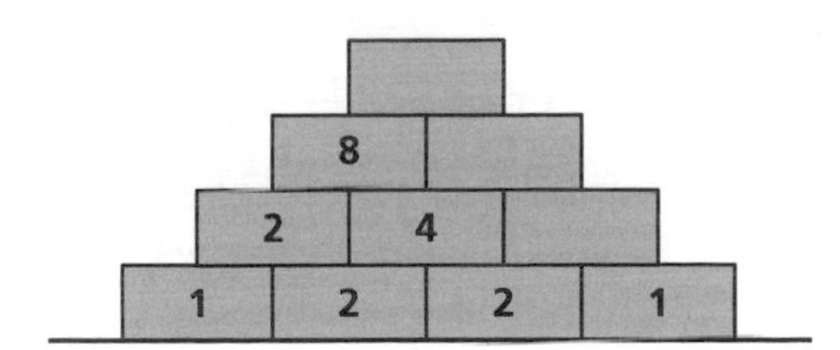

What number is on the top brick? ________

Test 32

Final score

$\overline{}$ 20 ____ %

For all of the questions in this test, do the calculation entirely in your head with no written 'working' and just write down the answer.

1 Write in figures the number which is eleven less than one hundred and two. ________

2 Birthday candles are sold in packs of 10 and cost £2.10 per pack. Kevin wants to buy 60 candles. How much will he pay?
£________

3 Find the value of $3^2 + 2^2$ ________

4 What is the area of a rectangle measuring 12.5 cm by 4 cm? ________ cm^2

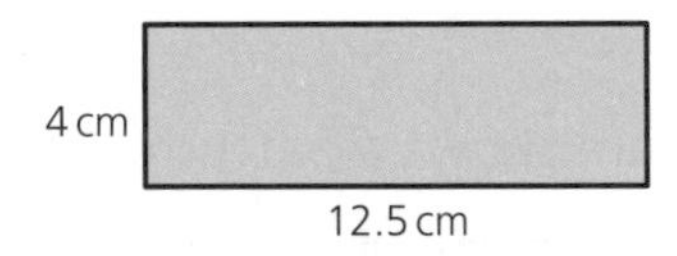

5 A 500 g pack of dog biscuits contains 140 biscuits. How many biscuits would there be in a 250 g box of dog biscuits? ________

6 Dr Graham buys a motorhome which normally costs £30 000, but the price is reduced by 10%.

What does Dr Graham pay? £________

7 George came second in a donkey race with a time of 5 minutes and 14 seconds. Mike came first with a time of 5 minutes and 9 seconds. By how many seconds did Mike beat George?

________ seconds

8 What is $6 \times 3 \times 5$? ________

9 Mrs Gray buys 4 sacks of potatoes costing £4.50 each. How much change does she receive from a £20 note?

£________

10 What is a quarter of 52? ________

11 Alastair is thinking of two integers (whole numbers). The sum of the integers is 40 and the difference between them is 4

What is the larger number? ________

12 Siobhan has the five number cards below.

What is the largest multiple of 3 she can make by placing all five cards side by side?

____

13 The Venn diagram shows some information about children in a class.

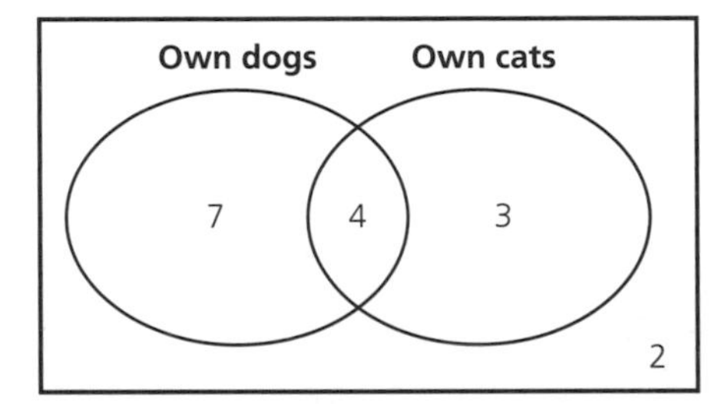

How many children in the class own either a dog or a cat but not both? ________

14 The **range** of the scores in set A is 5 (9 – 4).

What is the range of the scores in set B?

____

Set A: 9 6 9 4 8 7 7

Set B: 8 3 5 9 6 4 10

15 Which number between 80 and 90 is divisible exactly by 11? ________

16 What is the smallest number of British coins needed to pay exactly £4.88? ________

17 Talia has a collection of china figures. She gives $\frac{1}{3}$ of them to her little brother. She now has 8 figures. How many did she give to her brother? ________

18 How many of the following calculations give a result with a units digit 1? ________

$9 \times 19 \quad 99 \div 9 \quad 193 - 94 \quad 49 + 90$

19 Dora's birthday is 29 April and Jamie's is 7 days later. What date is Jamie's birthday?

____

20 What is the sum of the numbers 31 to 34 inclusive? ________

Test 33

Final score

$\overline{20}$ ___ %

For all of the questions in this test, do the calculation entirely in your head with no written 'working' and just write down the answer.

In questions 1 to 10 you are reminded of 10 useful strategies which may help you in later questions.

1 25×21 ________

✱ Take easier steps when you multiply

2 $22\,\text{cm} \times 8$ ________

✱ Double or halve

3 $170 \div 5$ ________

✱ Use your 10s and 2s

4 $79 + 69$ ________

✱ Take easier steps when you add

5 $196 \div 14$ ________

✱ Use factors to divide

6 1.2×1.2 ________

✱ Use known facts

7 $\frac{4}{7}$ of 28 ________

✱ Use a step-by-step approach

8 $47 + 38 + 53$ ________

✱ Group when you add

9 71×6 ________

✱ Approximate the result

10 23×7 ________

✱ Check using the units digit of the result

11 Olivia has a packet of 24 mints which she shares with Ruby and Jack. Ruby eats 6 mints, Jack eats 9 mints and Olivia eats the rest. How many mints does Olivia eat?

____

12 Emily has £7.40 in her purse. She hands over a £5 note to pay for a book and puts the change in her purse without checking. At home, she finds that she has £3.50

What was the price of the book?

£________

13 What is the cost of 6 bags of crisps priced at 49p each?

£________

14 Oliver left school at 15:45 and got home at 16:13. How long was his journey home?

________ minutes

15 What is the largest number, smaller than 50, which is a multiple of 3? ________

16 Given that $102 \times 28 = 2856$, what is 51×56?

____

17 By how much is 7×14 greater than 4×17?

____

18 What fraction of this shape is white?

____

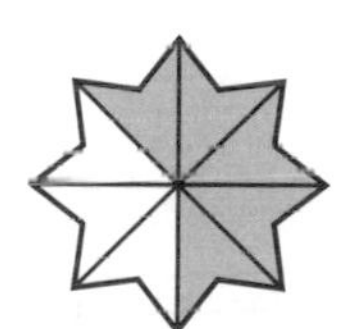

19 Brydon is thinking of a number and gives the following clues:

"My number is:

- odd
- between 30 and 50

The digits add to 7"

What number is Brydon thinking of? ________

20 On this tower of bricks, the number on each brick is the sum of the numbers on the two bricks supporting it.

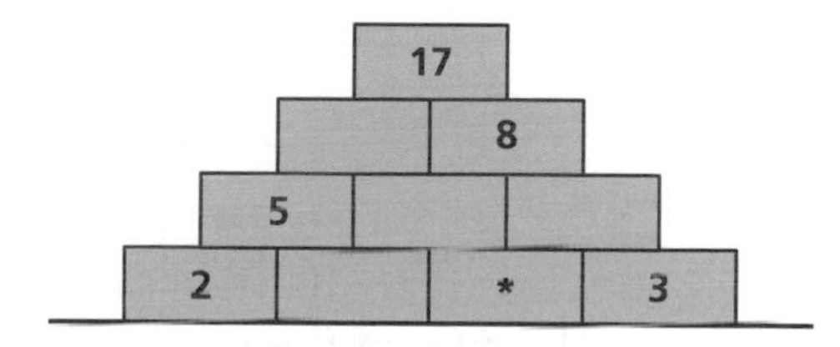

What number is on the brick marked with a star (*)? ________

Test 34

Final score

$\overline{}$
20 ____ %

For all of the questions in this test, do the calculation entirely in your head with no written 'working' and just write down the answer.

1 Write in figures the number which is forty-six less than ninety-two. ________

2 Small pork pies are sold in packs of 4 and cost £1.40 per pack. Grace buys 12 pork pies for the hockey team. What will be the total cost? £________

3 Find the value of $8^2 + 2^2$ ________

4 What is the perimeter of a rectangle measuring 17 cm by 4 cm? ________ cm

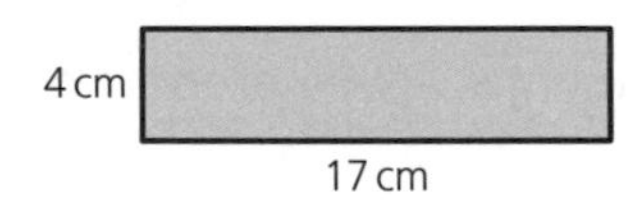

5 A pudding for 4 people requires 300 ml of milk. Jessica wants to make the same pudding for 6 people. How much milk should she use? ________ ml

6 A shirt normally costs £30

The price is reduced by 10%. What is the new price?

£________

7 Dylan finished second in a hopping race with a time of 2 minutes 3 seconds. Jasmine beat him by 9 seconds. What was Jasmine's time?

______ minute(s) ______ seconds

8 What is $3 \times 5 \times 12$? ________

9 Phil buys 2 DVDs costing £7.55 each. How much change does he receive from a £20 note?

£________

10 What is an eighth of 400? ________

11 Karen is thinking of two integers (whole numbers) both less than 10. Their product is 42. What is the sum of the numbers?

____

12 Henry has the five number cards below.

What is the largest 4-digit even number he can make by placing four of the cards side by side? ________

13 The Carroll diagram shows some information about pupils in Year 5

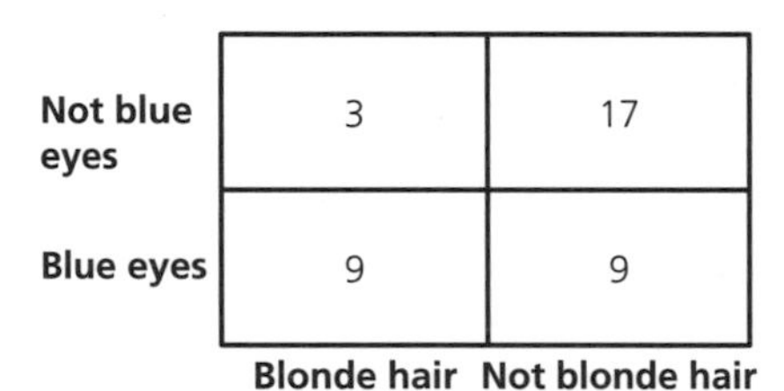

	Blonde hair	Not blonde hair
Not blue eyes	3	17
Blue eyes	9	9

How many pupils have neither blue eyes nor blonde hair? ________

14 What number is exactly half way between 20 and 52? ________

15 Which number between 60 and 70 is divisible exactly by 9? ________

16 What is the smallest number of British coins needed to pay exactly £2.27? ________

17 A pet shop has 36 pets for sale. A third are hamsters, a quarter are guinea pigs and the rest are rabbits. How many rabbits are there?

____

18 How many of the following calculations give a result with a units digit 7?

____

19×13 $54 \div 9$ $43 - 26$ $18 + 49$

19 On 31 March, Eli's age was recorded as 11 years 6 months. In which month is his birthday?

____

20 What is the sum of the numbers 29 to 31 inclusive?

____

Test 35

Final score

$\overline{20}$ ___ %

For all of the questions in this test, do the calculation entirely in your head with no written 'working' and just write down the answer.

In questions 1 to 10 you are reminded of 10 useful strategies which may help you in later questions.

1 119×3 ________

* Try 120×3 and then subtract 3

2 $240 \div 8$ ________

* Divide by 2, by 2 again and then by 2 again

3 43×5 ________

* Multiply by 10 then divide by 2

4 $400 - 78$ ________

* Subtract 80 then add 2

5 $102 \div 6$ ________

* Divide by 2, then by 3

6 $1.2 \div 3$ ________

* You know that $12 \div 3 = 4$

7 $\frac{2}{3}$ of 75 ________

* Find $\frac{1}{3}$ then multiply by 2

8 $13 + 45 + 87$ ________

* $13 + 87 = 100$ first

9 299×2 ________

* The result must be about 600

10 7×31 ________

* The units digit of the result is 7

11 Elise had forty 2p coins. She spent 48 pence on a biscuit. How many coins does she have left?

____

12 There were 33 children at an ice-rink. One-third of them were boys. How many girls were there?

____

13 What is the cost of 9 boxes of cat food priced at £10 for 3 boxes?

£________

14 Isabella left home at 07:45 and reached school at 08:20. How long did the journey take her?

________ minutes

15 What is the smallest number, greater than 40, which divides exactly by 3?

____

16 Given that $1200 \div 48 = 25$, what is $1200 \div 24$?

____

17 By how much is 9×16 greater than 6×19?

____

18 The diagram shows a pattern of tiles.

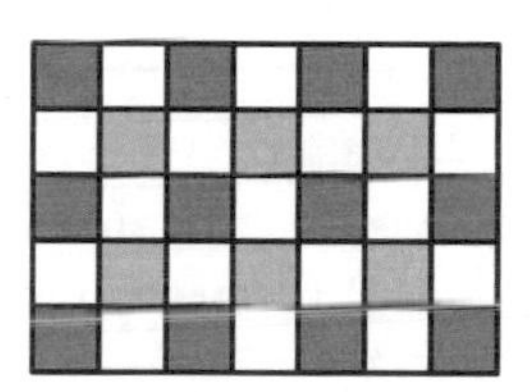

What fraction of the tiles is blue?

____

19 Noel is thinking of a number and gives the following clues:

"My number is:

- between 25 and 55
- 1 more than a prime number
- a multiple of 10"

What number is Noel thinking of? ________

20 On this tower of bricks, the number on each brick is the **product** of the numbers on the two bricks supporting it.

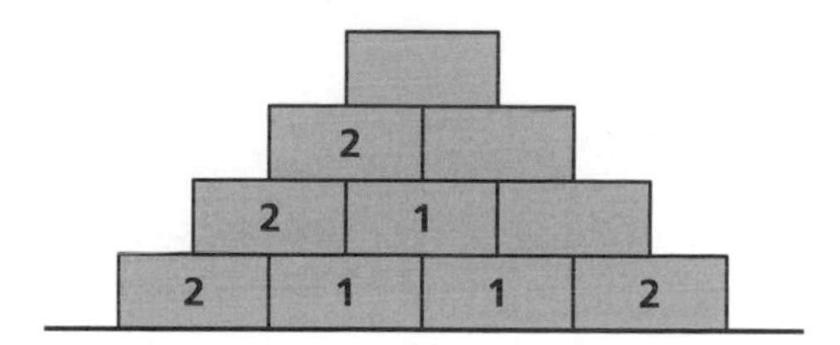

What number is on the top brick? ________

Test 36

Final score

$\overline{}$ 20 ____ %

For all of the questions in this test, do the calculation entirely in your head with no written 'working' and just write down the answer.

1 Write in figures the number which is sixteen more than one hundred and ninety-four.

____

2 Cup-cakes are sold in packs of 6 and cost £1.90 per pack. Yusif wants to buy 30 cup-cakes for his party. How much will he pay?

£__________

3 Find the value of $5^2 - 4^2$ __________

4 What is the area of a rectangle measuring 14 cm by 3.5 cm? __________ cm^2

3.5 cm

14 cm

5 A 50 g bag of sweets contains 20 sweets. How many sweets would there be in a 150 g bag of the same sweets? __________

6 Nathan buys a DVD which normally costs £10, but the price is reduced by 15%.

What does Nathan pay? £__________

7 Jack and Jill won a three-legged race with a time of 1 minute and 47 seconds, beating Janet and John by 19 seconds. What was the time taken by Janet and John?

________ minutes ________ seconds

8 What is $2 \times 9 \times 5$? __________

9 Miss Pringle buys 5 packets of crisps priced at 79p each. How much change does she receive from a £10 note?

£__________

10 What is a third of 66? __________

11 Ethan is thinking of two integers (whole numbers). The product of the integers is 20 and the difference between them is 1

What is the smaller number? __________

12 Julian has the five number cards below.

What is the largest multiple of 5 he can make by placing all five cards side by side?

____

13 The Venn diagram shows some information about children in a school.

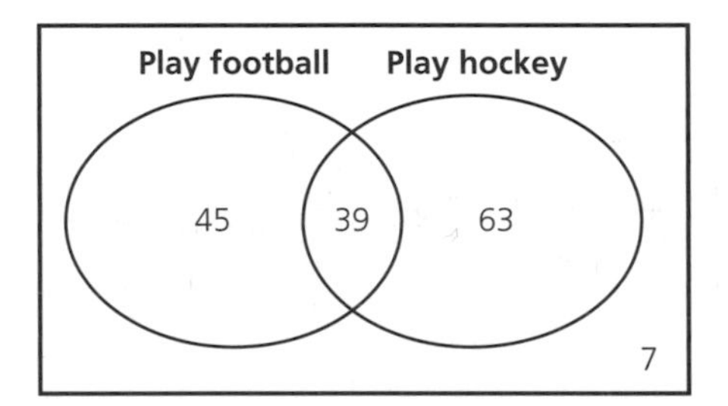

How many children are in the school?

____

14 The **range** of the scores in set A is 6 (9 – 3).

What is the range of the scores in set B?

____

Set A: 9 6 9 3 8 7 7

Set B: 7 6 6 9 6 7 10

15 Which number between 40 and 50 is divisible exactly by 9? __________

16 What is the smallest number of British coins needed to pay exactly £1.73? __________

17 Ava has a collection of bears. She gives $\frac{1}{4}$ of them to a charity shop. The charity shop received 12 bears. How many bears did Ava have in her collection? __________

18 How many of the following calculations give a result with a units digit 0? __________

4×15 $\quad 60 \div 3$ $\quad 189 - 99$ $\quad 37 + 73$

19 Summer's birthday is 29 August and Hannah's is 5 days later. What date is Hannah's birthday? ____________________

20 What is the sum of the numbers 49 to 51 inclusive? __________

Test 37

Final score

$\overline{20}$ ___ %

For all of the questions in this test, do the calculation entirely in your head with no written 'working' and just write down the answer.

In questions 1 to 10 you are reminded of 10 useful strategies which may help you in later questions.

1 41×31 ________

✱ Take easier steps when you multiply

2 33×8 ________

✱ Double or halve

3 $140 \div 5$ ________

✱ Use your 10s and 2s

4 $88 + 49$ ________

✱ Take easier steps when you add

5 $192 \div 16$ ________

✱ Use factors to divide

6 1.1×9 ________

✱ Use known facts

7 $\frac{3}{8}$ of 56 ________

✱ Use a step-by-step approach

8 $65 + 29 + 55$ ________

✱ Group when you add

9 69×3 ________

✱ Approximate the result

10 41×6 ________

✱ Check using the units digit of the result

11 Amber has a packet of 20 sweets which she shares with Rylan and Jules. They take turns to pick a sweet, with Amber going third, until the packet is empty. How many sweets does Rylan get? ________

12 George has £5.27 in his pocket. He takes three £1 coins to pay for a magazine and puts the change in his pocket. He now has £2.32. How much was the magazine?

£________

13 What is the cost of 8 bags of crisps priced at 53p each?

£________

14 Arthur left school at 16:37 and got home at 17:01. How long was his journey home?

________ minutes

15 What is the largest number, less than 40, which is a multiple of 3? ________

16 Given that $88 \times 90 = 7920$, what is 44×90?

____

17 By how much is 8×14 greater than 4×18?

____

18 What fraction (in its simplest form) of this shape is green?

____

19 Callum is thinking of a number and gives the following clues:

"My number is:

- odd
- between 30 and 50

The digits add to 11"

What number is Callum thinking of? ________

20 On this tower of bricks, the number on each brick is the sum of the numbers on the two bricks supporting it.

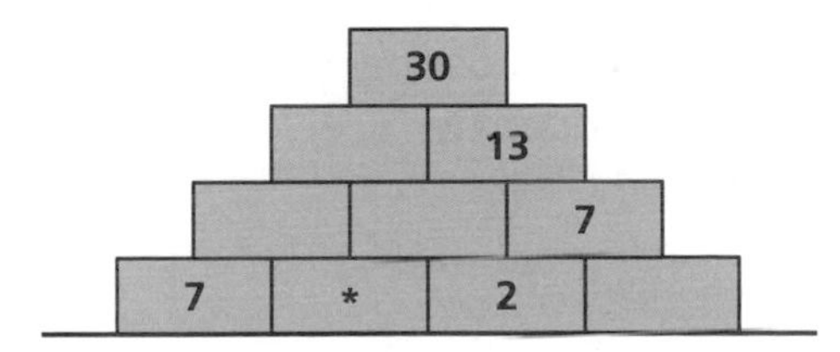

What number is on the brick marked with a star (*)?

____

Test 38

Final score

$\overline{}$ 20 ___ %

For all of the questions in this test, do the calculation entirely in your head with no written 'working' and just write down the answer.

1 Write in figures the number which is fourteen more than one hundred and eighteen.

___

2 Sausages are sold in packs of 6 and cost £1.90 per pack. Jack buys 30 sausages for a camping trip. What will be the total cost?
£___

3 Find the value of $11^2 + 7^2$ ___

4 What is the perimeter of a rectangle measuring 12 cm by 7 cm? ___ cm

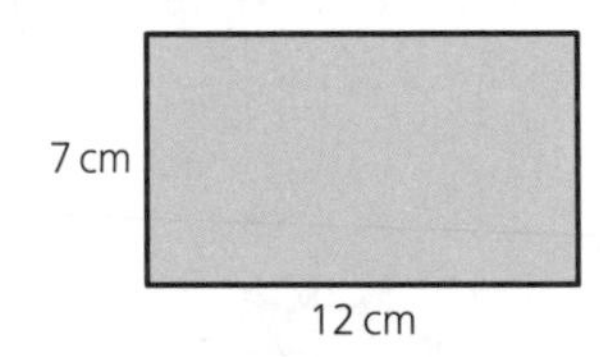

5 A dessert for 8 people requires 200 ml of milk. Georgia wants to make the same dessert for 12 people. How much milk should she use?
___ ml

6 A sweater normally costs £36

The price is reduced by a quarter. What is the new price? £___

7 Elaine finished second in a swimming race with a time of 1 minute 5 seconds. Morgan beat her by 7 seconds. What was Morgan's time?

___ seconds

8 What is $2 \times 3 \times 7$? ___

9 Chris buys 3 CDs costing £8.50 each. How much change does he receive from two £20 notes? £___

10 What is a seventh of 35? ___

11 Danny is thinking of two integers (whole numbers) both less than 10. Their product is 48. What is the sum of the numbers?

___

12 Lois has the five number cards below.

What is the largest 4-digit even number she can make by placing four of the cards side by side?

___

13 The Carroll diagram shows some information about pupils in a class.

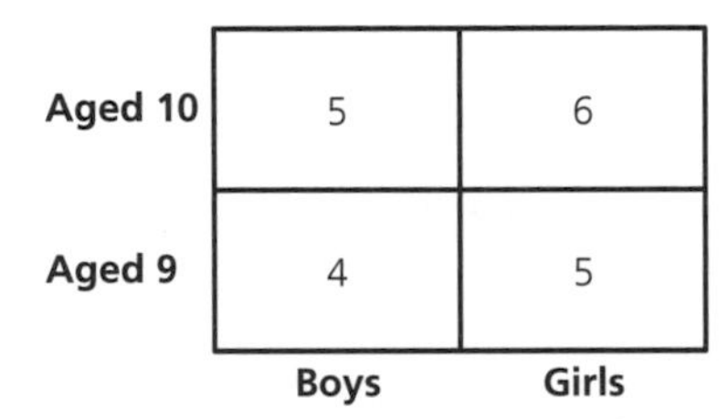

What percentage of the pupils in the class is boys?

___ %

14 What number is exactly half way between 30 and 64? ___

15 Which number between 80 and 90 is divisible exactly by 7? ___

16 What is the smallest number of British coins needed to pay exactly £3.48? ___

17 A garage has 15 cars for sale. Two-thirds are second-hand and the rest are new. How many new cars are there? ___

18 How many of the following calculations give a result with a units digit 0?

___

24×15 $60 \div 2$ $303 - 43$ $47 + 33$

19 On 31 December, Paul's age was recorded as 9 years 10 months. In which month is his birthday?

___

20 What is the sum of the numbers 43 to 45 inclusive?

___

Test 39

Final score

$\overline{20}$ ___ %

For all of the questions in this test, do the calculation entirely in your head with no written 'working' and just write down the answer.

In questions 1 to 10 you are reminded of 10 useful strategies which may help you in later questions.

1 49×11 __________

* Try 49×10 and then add 49

2 123×4 __________

* Multiply by 2 and then by 2 again

3 $130 \div 5$ __________

* Multiply by 10 then divide by 2

4 $119 + 47$ __________

* Add 50 then subtract 3

5 $360 \div 15$ __________

* Divide by 3 then by 5

6 0.8×5 __________

* You know that $8 \times 5 = 40$

7 $\frac{3}{5}$ of 55 __________

* Find $\frac{1}{5}$ then multiply by 3

8 $109 + 56 + 11$ __________

* $109 + 11 = 120$ first

9 39×9 __________

* The result must be about 400

10 44×5 __________

* The units digit of the result is 0

11 Jake has a packet of 12 biscuits which he shares equally with Luke and Matthew. What fraction of the biscuits does Jake eat?

____

12 Millie has £6.47 in her purse. She hands over a £2 coin to pay for a chocolate bar and puts the change in her purse without checking. At home, she finds that she has £5.65.

What was the price of the chocolate bar?

__________ pence

13 What is the cost of 4 bags of cement priced at £2.49 each?

£__________

14 Oscar left school at 15:50 and took 43 minutes to walk home. At what time did he reach home?

__________:__________

15 What is the largest number, less than 100, which is a multiple of 3? __________

16 Given that $28 \times 32 = 896$, what is 27×32?

____

17 By how much is 9×12 greater than 2×19?

____

18 What percentage of this rectangle is purple?

__________ %

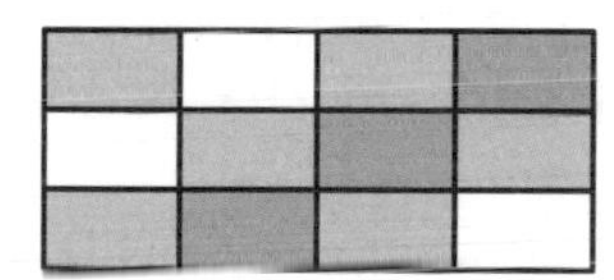

19 Archie is thinking of a number and gives the following clues:

"My number is:

- even
- between 40 and 60

The digits add to 9"

What number is Archie thinking of? __________

20 On this tower of bricks, the number on each brick is the **product** of the numbers on the two bricks supporting it.

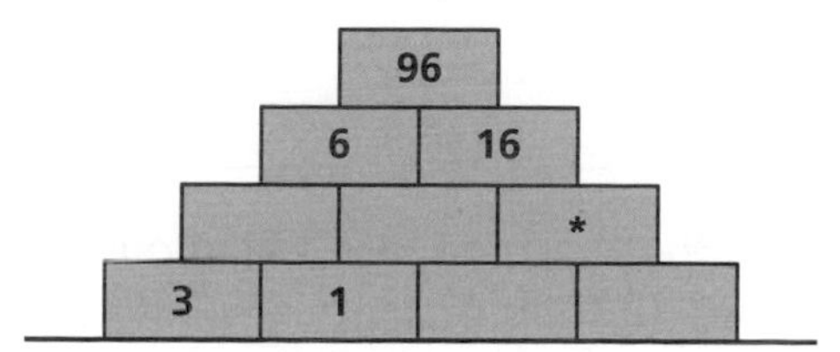

What number is on the brick marked with a star (*)? __________

Test 40

Final score

/20 ____ %

For all of the questions in this test, do the calculation entirely in your head with no written 'working' and just write down the answer.

1 Write in figures the number which is sixty-seven less than one hundred and twenty.

____

2 Yoghurts are sold in packs of 4 and cost £2.65 per pack. Grace buys 8 yoghurts. What will be the total cost? £__________

3 Find the value of $8^2 - 2^2$ __________

4 What is the perimeter of a square of side 7.5 cm? __________ cm

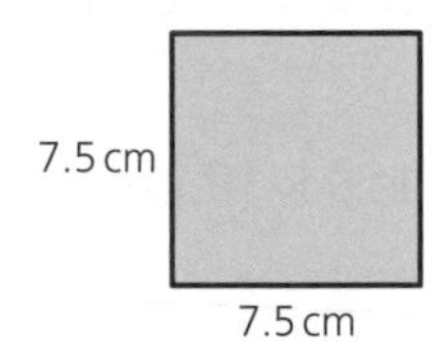

5 A pudding for 6 people requires 300 ml of milk. Lily wants to make the same pudding for 4 people. How much milk should she use?

__________ ml

6 A new violin normally costs £350

The price is reduced by 10%. What is the new price? £__________

7 Evie finished first in a speed quiz with a time of 7 minutes 38 seconds. Lucy finished 37 seconds later. What was Lucy's time?

_______ minutes _______ seconds

8 What is $6 \times 6 \times 5$? __________

9 Jayden buys 3 magazines costing £2.99 each. How much change does he receive from a £10 note?

£__________

10 What is a sixth of 150? __________

11 Adam is thinking of three prime numbers all less than 10. Their product is 30. What is the sum of the numbers?

____

12 Daisy has the six number cards below.

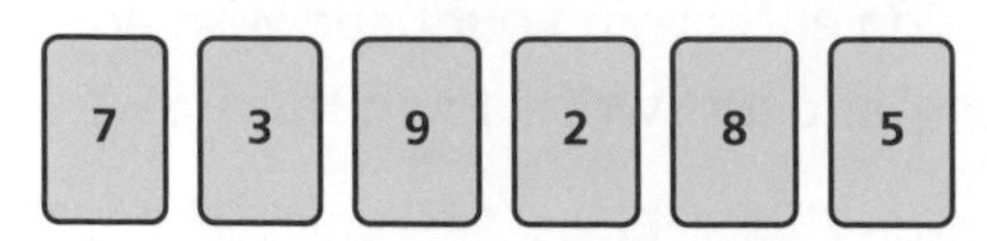

What is the number closest to 1000 that she can make by placing some of the cards side by side? __________

13 The Carroll diagram shows some information about pupils in Year 5

	Have sisters	Do not have sisters
Do not have brothers	7	8
Have brothers	6	10

How many pupils have both brothers and sisters? __________

14 What number is exactly half way between 40 and 64? __________

15 Which number between 80 and 90 is divisible exactly by 12? __________

16 What is the smallest number of British coins needed to pay exactly £4.44? __________

17 A small farm has 64 animals. Half are sheep, a quarter are cows and the rest are pigs. How many pigs are there?

____

18 How many of the following calculations give a result with a units digit 5?

____

15×13 $\quad 55 \div 5$ $\quad 42 - 27$ $\quad 19 + 43$

19 On 31 December, Phoebe's age was recorded as 10 years 3 months. In which month is her birthday?

____

20 What is the sum of the numbers 23 to 25 inclusive?

____

Test 41

Final score

$\overline{20}$ ___ %

For all of the questions in this test, do the calculation entirely in your head with no written 'working' and just write down the answer.

In questions 1 to 10 you are reminded of 10 useful strategies which may help you in later questions.

1 59×30 __________

✱ Take easier steps when you multiply

2 53×4 __________

✱ Double or halve

3 28×5 __________

✱ Use your 10s and 2s

4 $48 + 89$ __________

✱ Take easier steps when you add

5 $384 \div 12$ __________

✱ Use factors to divide

6 1.2×5 __________

✱ Use known facts

7 $\frac{4}{5}$ of 40 __________

✱ Use a step-by-step approach

8 $43 + 29 + 51 + 7$ __________

✱ Group when you add

9 99×5 __________

✱ Approximate the result

10 53×7 __________

✱ Check using the units digit of the result

11 Freya has a bag of plums which she shares equally with Abigail and Poppy. When they have each eaten 5 plums, there is 1 plum left in the bag. How many plums were in the bag at the start? __________

12 Edward has £7.20 in his pocket. He takes four £1 coins to pay for lunch and puts the change in his pocket. He now has £3.45

How much was his lunch?

£__________

13 What is the cost of 5 cereal bars priced at 62p each? £__________

14 Leo left home at 10.45 a.m. to walk to the zoo. He reached the zoo at 11.25 a.m.

How long did the walk take him?

__________ minutes

15 What is the smallest number, greater than 50, which is a multiple of 4? __________

16 Given that $56 \times 48 = 2688$, what is 28×48?

____

17 By how much is 8×17 greater than 7×18?

____

18 What fraction of this shape is red?

____

19 Muhammad is thinking of a number and gives the following clues:

"My number is:

- even
- between 20 and 30

The digits add to 8"

What number is Muhammad thinking of?

____

20 On this tower of bricks, the number on each brick is the sum of the numbers on the two bricks supporting it.

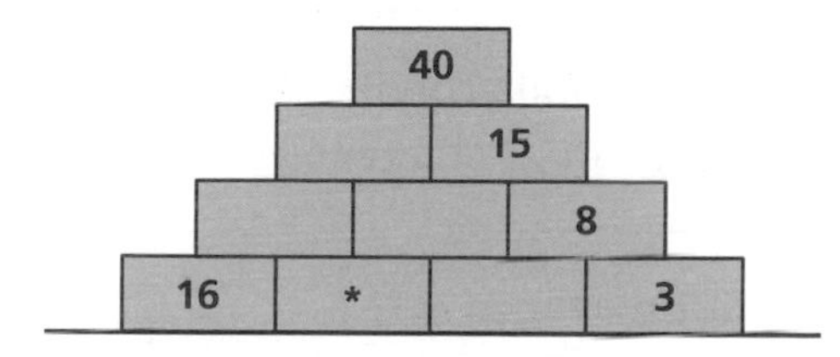

What number is on the brick marked with a star (*)?

____

Test 42

Final score

$\overline{}$ 20 ____ %

For all of the questions in this test, do the calculation entirely in your head with no written 'working' and just write down the answer.

1 Write in figures the number which is twenty-five less than eighty-four. __________

2 Sausage rolls are sold in packs of 3 and cost £2.10 per pack. Jack buys 24 sausage rolls for a party. What will be the total cost? £__________

3 Find the value of $5^2 + 5$ __________

4 What is the perimeter of a rectangle measuring 5.5 cm by 7.5 cm? __________ cm

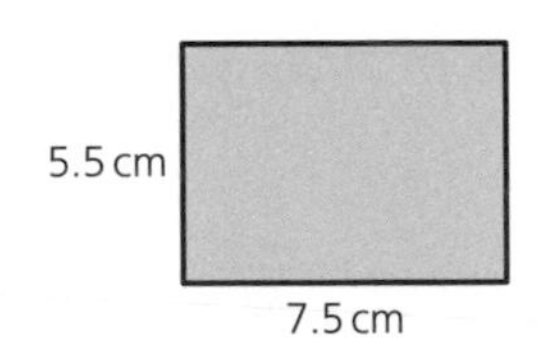

5 A pudding for 4 people requires 220 ml of milk. Erin wants to make the same pudding for 10 people. How much milk should she use? __________ ml

6 A scarf normally costs £25

The price is reduced by 10%. What is the new price?

£__________

7 Noah finished second in a swimming race with a time of 3 minutes 4 seconds. Cameron beat him by 11 seconds. What was Cameron's time?

________ minutes ________ seconds

8 What is $4 \times 5 \times 6$? __________

9 Emma buys 3 dolls costing £19.50 each. How much change does she receive from three £20 notes? £__________

10 What is a quarter of £9? £__________

11 Alex is thinking of two even integers (whole numbers) both less than 10. Their sum is 14. What is the product of the numbers?

____

12 Molly has the five number cards below.

What is the number nearest to 200 she can make by placing some of the cards side by side? __________

13 The Carroll diagram shows some information about members of a sports club.

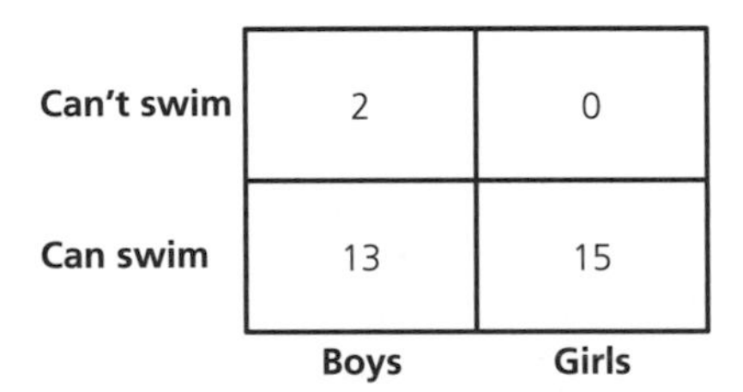

What fraction (in its simplest form) of the members can't swim?

____

14 What number is exactly half way between 18 and 28? __________

15 Which number between 100 and 105 is divisible exactly by 4? __________

16 What is the smallest number of British coins needed to pay exactly £1.99? __________

17 A marina has 12 yachts for sale. A third are second-hand and the rest are new. How many new yachts are there?

____

18 How many of the following calculations give a result with a units digit 4?

____

14×10 $\quad 80 \div 2$ $\quad 444 - 44$ $\quad 21 + 33$

19 On 31 March, Isla's age was recorded as 9 years 11 months. In which month is her birthday? __________

20 What is the sum of the numbers 1 to 5 inclusive?

____

Test 43

Final score

$\overline{20}$ ____ %

For all of the questions in this test, do the calculation entirely in your head with no written 'working' and just write down the answer.

In questions 1 to 10 you are reminded of 10 useful strategies which may help you in later questions.

1 69×4 ________

✱ Try 70×4 and then subtract 4

2 $640 \div 8$ ________

✱ Divide by 2, by 2 again and then by 2 again

3 44×5 ________

✱ Multiply by 10 then divide by 2

4 $700 - 358$ ________

✱ Subtract 360 then add 2

5 $330 \div 15$ ________

✱ Divide by 3, then by 5

6 0.9×8 ________

✱ You know that $9 \times 8 = 72$

7 $\frac{3}{4}$ of 64 ________

✱ Find $\frac{1}{4}$ then multiply by 3

8 $13 + 25 + 27 + 15$ ________

✱ $13 + 27$ and $25 + 15$

9 98×5 ________

✱ The result must be about 500

10 7×47 ________

✱ The units digit of the result is 9

11 Isla had 16 beetles in a box but 8 escaped. Isla recaptures half of the escaped beetles and puts them back in the box. How many beetles are in the box now ? ________

12 There are 40 musicians in the school orchestra. 40% of them play violins. How many violinists are there? ________

13 Bottles of juice are priced at 85p each but there is a special offer!

'Buy two, get one free'.

Owen wants 3 bottles of juice. How much will he pay? £________

14 Nathan started jogging at 15:50 and stopped at 16:15. For how long was he jogging? ________ minutes

15 What is the smallest number, greater than 60, which divides exactly by 9? ________

16 Given that $1748 \div 46 = 38$, what is $1748 \div 92$? ________

17 By how much is 7×16 greater than 6×17? ________

18 The diagram shows a pattern of tiles.

How many more white tiles are there than orange tiles? ________ more

19 Imogen is thinking of a number and gives the following clues:

"My number is:

- between 20 and 35
- 1 more than a multiple of 3
- prime."

What number is Imogen thinking of? ________

20 On this tower of bricks, the number on each brick is the **product** of the numbers on the two bricks supporting it.

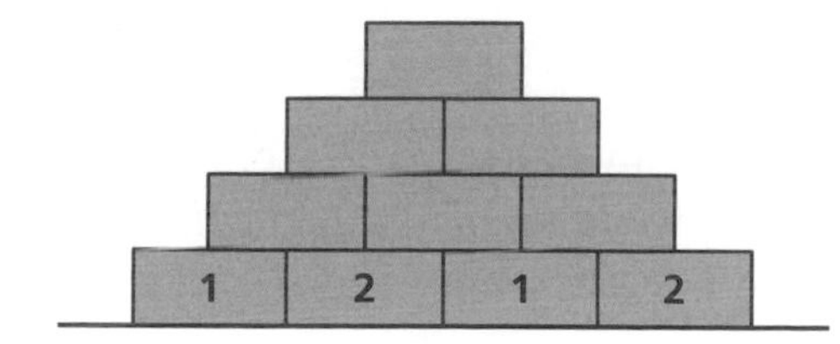

What number is on the top brick? ________

Test 44

Final score

/20 ___ %

For all of the questions in this test, do the calculation entirely in your head with no written 'working' and just write down the answer.

1 Write in figures the number which is seventeen less than one hundred and five.

____

2 Christmas cards are sold in packs of 10 and cost £2.35 per pack. Scarlett wants to buy 40 cards. How much will she pay?

£___________

3 Find the value of $1^2 + 7^2$ ___________

4 What is the area of a rectangle measuring 2.5 cm by 12 cm? ___________ cm²

2.5 cm

12 cm

5 A 150 g pack of biscuits contains 30 biscuits. How many biscuits would there be in a 250 g pack of biscuits? ___________

6 Michael buys a bicycle which normally costs £240, but the price is reduced by a third.

What does Michael pay? £___________

7 Toby came second in a cricket ball throwing competition with a throw of 39.78 metres. Louis's winning throw was 45 centimetres further. How far did Louis throw the ball?

___________ m

8 What is $5 \times 19 \times 2$? ___________

9 Freddie buys 5 boxes of tissues costing £1.99 each. How much change does he receive from a £20 note?

£___________

10 What is a quarter of 10? ___________

11 Alastair is thinking of two integers (whole numbers). The sum of the integers is 30 and the difference between them is 2

What is the larger number? ___________

12 Finlay has the five number cards below.

What is the largest square number he can make by placing two of the cards side by side? ___________

13 The Venn diagram shows some information about children in a class.

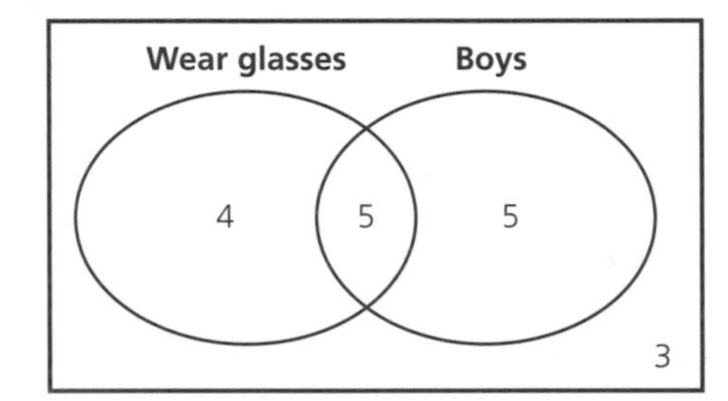

How many girls are in the class? ___________

14 The **range** of the scores in set A is 3 (9 – 6).

What is the range of the scores in set B?

____

Set A: 9 6 9 6 8 7 7

Set B: 6 7 9 9 6 7 9

15 Which number between 50 and 60 is divisible exactly by 13? ___________

16 What is the smallest number of British coins needed to pay exactly 85 pence? ___________

17 Leon has a collection of model cars. He gives $\frac{1}{4}$ of them to his little brother. Leon now has 12 model cars. How many did he give to his brother? ___________

18 How many of the following calculations give a result with a units digit 6? ___________

4×14 $36 \div 6$ $120 - 94$ $37 + 29$

19 Leah's birthday is 3 November and Sophia's is 7 days earlier. What date is Sophia's birthday?

____

20 What is the sum of the numbers 100 to 104 inclusive? ___________

Test 45

Final score

$\overline{20}$ ___ %

For all of the questions in this test, do the calculation entirely in your head with no written 'working' and just write down the answer.

In questions 1 to 10 you are reminded of 10 useful strategies which may help you in later questions.

1 11×19 ________

✱ Take easier steps when you multiply

2 41×8 ________

✱ Double or halve

3 $180 \div 5$ ________

✱ Use your 10s and 2s

4 $68 + 39$ ________

✱ Take easier steps when you add

5 $288 \div 18$ ________

✱ Use factors to divide

6 0.8×7 ________

✱ Use known facts

7 $\frac{5}{6}$ of 36 ________

✱ Use a step-by-step approach

8 $28 + 27 + 52 + 5$ ________

✱ Group when you add

9 59×4 ________

✱ Approximate the result

10 17×11 ________

✱ Check using the units digit of the result

11 Elizabeth has a packet of 26 sweets which she shares with some friends. When Elizabeth and her friends have each received the same number of sweets there is just 1 sweet left. How many sweets did Elizabeth and her friends each get?

____

12 Harley has seven 20p coins in his pocket. He takes four of the coins to pay for a bag of crisps and receives 7p in change. How much money does Harley have left? ________ pence

13 What is the cost of 5 packets of chews priced at 49p each? £________

14 Reece left school at 15:48 and got home at 16:29. How long was his journey home?

________ minutes

15 What is the largest number, less than 1000, which is a multiple of 3? ________

16 Given that $65 \times 40 = 2600$, what is 66×40?

____

17 By how much is 6×14 greater than 4×16?

____

18 What fraction (in its simplest form) of this shape is purple?

____

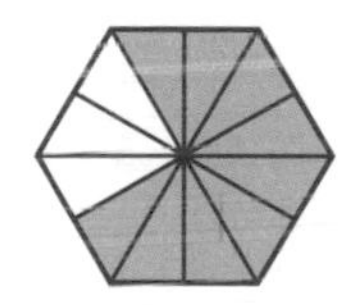

19 Kian is thinking of a number and gives the following clues:

"My number is:

- even
- between 20 and 40

The digits add to 5"

What number is Kian thinking of? ________

20 On this tower of bricks, the number on each brick is the sum of the numbers on the two bricks supporting it.

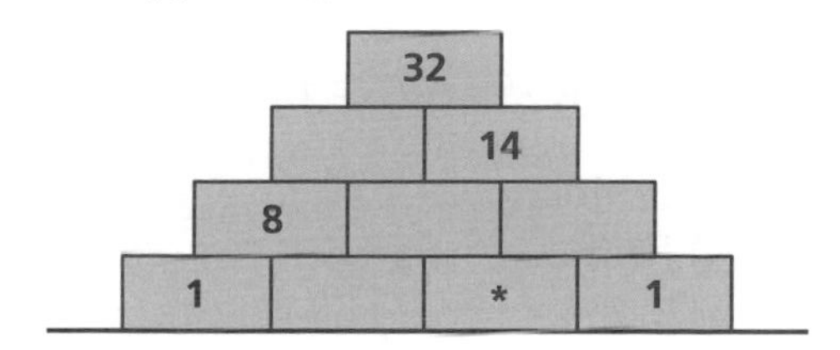

What number is on the brick marked with a star (*)? ________

Test 46

Final score

$\overline{}$ 20 ___ %

For all of the questions in this test, do the calculation entirely in your head with no written 'working' and just write down the answer.

1 Write in figures the number which is thirteen less than one hundred and one. __________

2 Muffins are sold in packs of 4 priced at £2.28 per pack. What is the cost of 1 muffin?
__________ pence

3 Find the value of $4^2 + 6^2$ __________

4 What is the perimeter of a rectangle measuring 12.5 cm by 6.5 cm? __________ cm

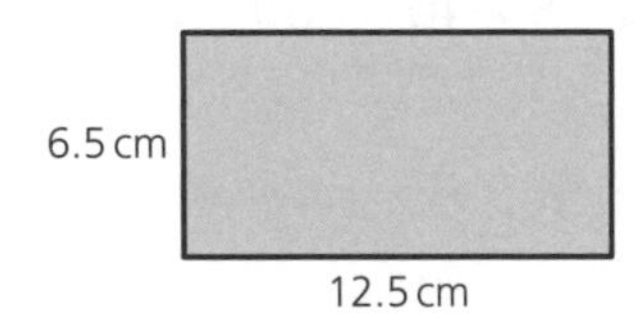

5 A sweet course for 4 people requires 150 ml of milk. Brooke wants to make the same sweet course for 12 people. How much milk should she use? __________ ml

6 A hoodie normally costs £28

The price is reduced by a quarter. What is the new price?

£__________

7 Matilda finished second in the long jump with a distance of 3.90 metres. Caitlin beat her by 7 centimetres. What was Caitlin's distance?
__________ m

8 What is $12 \times 5 \times 2$? __________

9 Kyle buys 4 CDs priced at £7.99 each. How much change does he receive from a £50 note?

£__________

10 What is a tenth of 102 cm? __________ cm

11 Brandon is thinking of two integers (whole numbers) both less than 10. Their product is 28. What is the sum of the numbers?

____

12 Keira has the six number cards below.

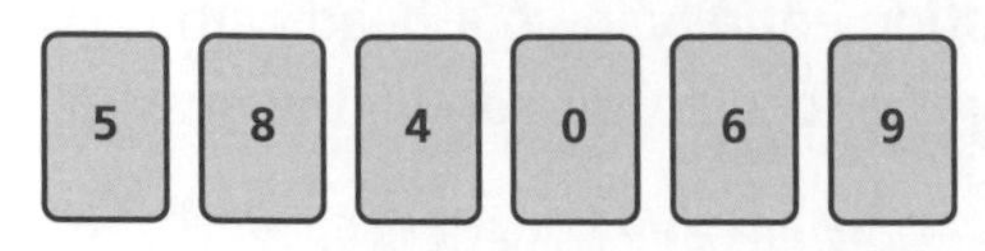

What is the number closest to 5000 she can make by placing four of the cards side by side? __________

13 The Carroll diagram shows some information about pupils in a class.

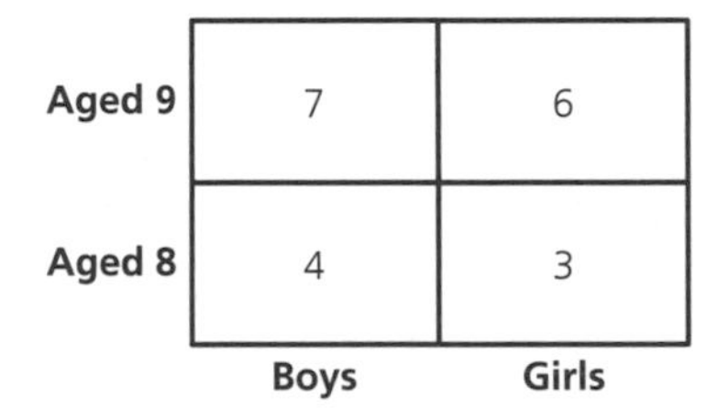

	Boys	Girls
Aged 9	7	6
Aged 8	4	3

What percentage of the pupils in the class is aged 9? __________ %

14 What number is exactly half way between 45 and 59? __________

15 Which number between 70 and 80 is divisible exactly by 9? __________

16 What is the smallest number of British coins needed to pay exactly £3.20? __________

17 A cycle shop has 24 cycles for sale. Three-quarters are mountain bikes and the rest are road bikes. How many road bikes are there?

____

18 How many of the following calculations give a result with a units digit 3?

____

27×109 $45 \div 15$ $102 - 29$ $28 + 35$

19 On 31 December, Alice's age was recorded as 9 years 8 months. In which month is her birthday?

____

20 What is the sum of the numbers 61 to 63 inclusive?

____

Test 47

Final score

$\overline{20}$ ___ %

For all of the questions in this test, do the calculation entirely in your head with no written 'working' and just write down the answer.

In questions 1 to 10 you are reminded of 10 useful strategies which may help you in later questions.

1 39×6 ________

* Try 40×6 and then subtract 6

2 $440 \div 8$ ________

* Divide by 2, by 2 again and then by 2 again

3 43×5 ________

* Multiply by 10 then divide by 2

4 $400 - 159$ ________

* Subtract 160 then add 1

5 $630 \div 15$ ________

* Divide by 3, then by 5

6 1.1×7 ________

* You know that $11 \times 7 = 77$

7 $\frac{2}{3}$ of 54 ________

* Find $\frac{1}{3}$ then multiply by 2

8 $14 + 28 + 32 + 26$ ________

* $14 + 26$ and $28 + 32$

9 49×5 ________

* The result must be about 250

10 8×15 ________

* The units digit of the result is 0

11 Zachary had 30 mice in a large cage but half of them escaped. Zachary recaptured two-thirds of the escaped mice and put them back in the cage. How many mice are in the cage now?

____

12 There are 25 members of the school choir. There are 3 more girls than boys. How many boys are there?

____

13 Cans of lemonade are priced at 80p each but there is a special offer!

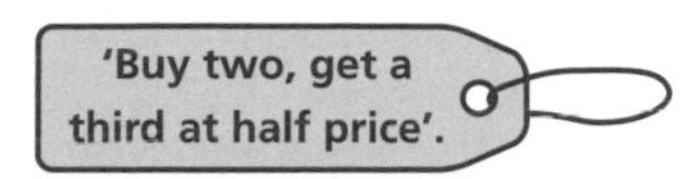

Kieran wants 6 cans of lemonade. How much will he pay? £________

14 Lola started her lunch break at 12.50 p.m. and started working again 45 minutes later. At what time did she start working? ________

15 What is the smallest number, greater than 45, which divides exactly by 6? ________

16 Given that $1044 \div 36 = 29$, what is $2088 \div 29$?

____

17 By how much is 9×15 greater than 5×19?

____

18 The diagram shows a pattern of tiles.

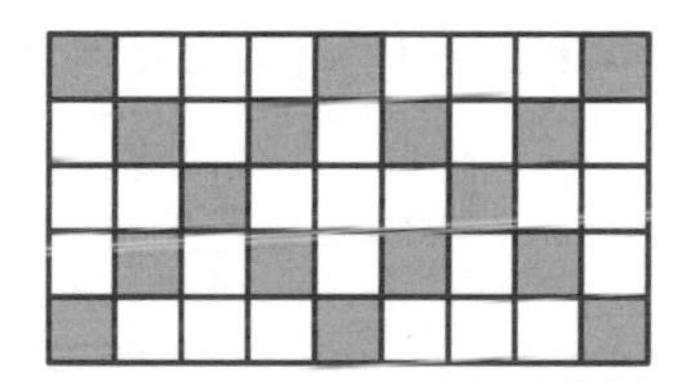

How many more white tiles are there than orange tiles? ________ more

19 Lilly is thinking of a number and gives the following clues:

"My number is:

- between 40 and 50
- 1 more than a multiple of 3
- prime."

What number is Lilly thinking of? ________

20 On this tower of bricks, the number on each brick is the **product** of the numbers on the two bricks supporting it.

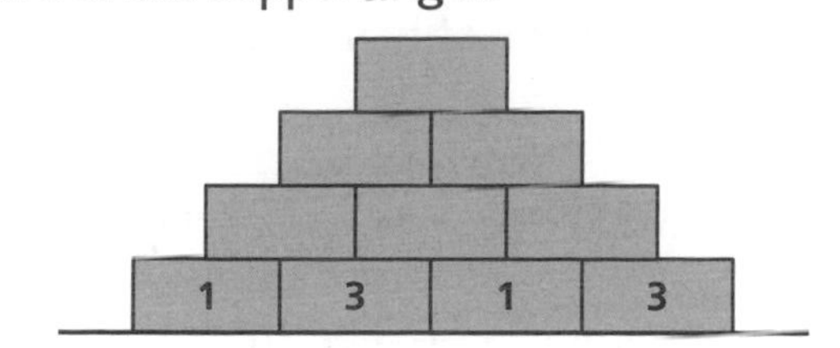

What number is on the top brick? ________

Test 48

Final score

$\overline{20}$ ___ %

For all of the questions in this test, do the calculation entirely in your head with no written 'working' and just write down the answer.

1 Write in figures the number which is twenty-four less than two hundred. ________

2 Trading cards are sold in sets of 36 and cost £6.50 per pack. Luca bought sets and has 180 cards. How much did Luca pay?

£________

3 Find the value of $3^2 + 7$ ________

4 What is the area of a rectangle measuring 1.5 cm by 20 cm?

________ cm²

1.5 cm

20 cm

5 A 380 g pack of shortbread fingers contains 20 fingers. What is the mass of 1 shortbread finger?

________ g

6 Lauren buys a book which normally costs £4.90, but is marked down to half price.

What does Lauren pay? £________

7 Ashton won a high jump competition with a best jump of 1.23 m, beating Bailey's best jump by 4 cm. What was Bailey's best jump?

________ m

8 What is $7 \times 9 \times 5$? ________

9 Sebastian buys 6 bread rolls costing 36 pence each. How much change does he receive from a £5 note?

£________

10 What is two-thirds of half a cake?

________ of a cake

11 Georgia is thinking of two integers (whole numbers). The product of the integers is 40 and the difference between them is 3

What is the larger number? ________

12 Gabriel has the seven number cards below.

What is the number closest to a million he can make by placing some (or all) of the cards side by side? ________

13 The Venn diagram shows some information about children in a class.

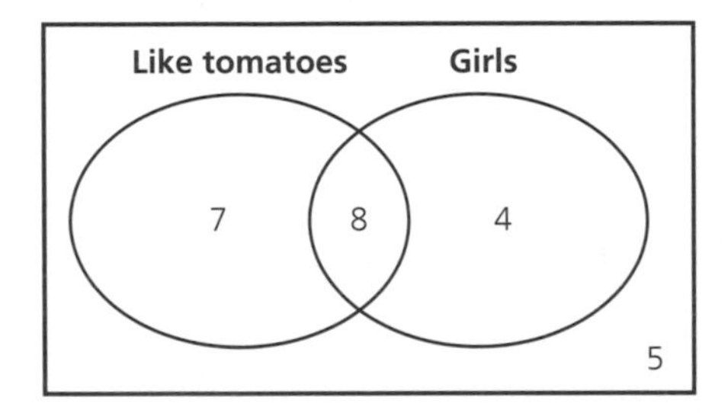

What fraction (in its simplest form) of the girls like tomatoes?

____

14 What is the mode of the test scores below?

____

9 6 9 6 8 7 7 6 7 9 9 6 7 9

15 Which number between 100 and 110 is divisible exactly by 7? ________

16 What is the smallest number of British coins needed to pay exactly £1.29? ________

17 Gracie has a collection of dolls. She gives a quarter of them to a charity shop and then gives her little sister 3. If Gracie now has 9 dolls, how many did she have originally?

____

18 How many of the following calculations give a result with a units digit 3? ________

$4 \times 13 \quad 42 \div 14 \quad 121 - 88 \quad 33 + 43$

19 Eleanor's birthday is 4 May and Bill's is 6 days earlier. What date is Bill's birthday?

____

20 What is the sum of the numbers 200 to 203 inclusive? ________

Test 49

Final score

$\overline{20}$ ___ %

For all of the questions in this test, do the calculation entirely in your head with no written 'working' and just write down the answer.

In questions 1 to 10 you are reminded of 10 useful strategies which may help you in later questions.

1 11×49 ________

✱ Take easier steps when you multiply

2 16×8 ________

✱ Double or halve

3 $190 \div 5$ ________

✱ Use your 10s and 2s

4 $147 + 59$ ________

✱ Take easier steps when you add

5 $240 \div 15$ ________

✱ Use factors to divide

6 1.2×1.1 ________

✱ Use known facts

7 $\frac{3}{5}$ of 120 ________

✱ Use a step-by-step approach

8 £3.40 + £1.95 + £1.60 £________

✱ Group when you add

9 101×9 ________

✱ Approximate the result

10 14×8 ________

✱ Check using the units digit of the result

11 Bethany bought a case of 15 cans of cola on Monday. She drinks 1 can each day. On which day of the week will she drink the last can?

____

12 Evan has 36 marbles. He loses 5 to Bradley and wins 13 from Elliot. How many marbles does he have now?

____

13 What is the cost of 7 DVDs priced at £10.50 each?

____

14 Madison attended a $2\frac{1}{2}$ hour concert which finished at 21:10. At what time did the concert start?

________ : ________

15 What is the smallest number, greater than 1000, which is a multiple of 3?

____

16 Given that $432 \times 15 = 6480$, what is 216×15?

____

17 By how much is 9×13 greater than 3×19?

____

18 What fraction of this shape is blue?

____

19 Amelie is thinking of a number and gives the following clues:

"My number is:

- odd
- a multiple of 7
- between 90 and 100"

What number is Amelie thinking of? ________

20 On this tower of bricks, the number on each brick is the sum of the numbers on the two bricks supporting it.

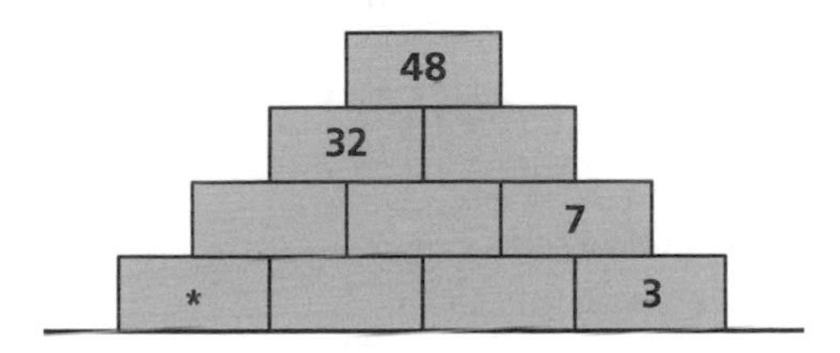

What number is on the brick marked with a star (*)? ________

Test 50

Final score

$\overline{20}$ ___ %

For all of the questions in this test, do the calculation entirely in your head with no written 'working' and just write down the answer.

1 Write in figures the number which is one hundred and four less than two hundred and one. __________

2 Muffins are sold in packs of 4 and cost £2.99 per pack. Annabel needs 16 muffins for a coffee morning. What will be the total cost?
£__________

3 Find the value of $12^2 + 11^2$ __________

4 What is the perimeter of a rectangle measuring 17.5 cm by 3.5 cm? __________ cm

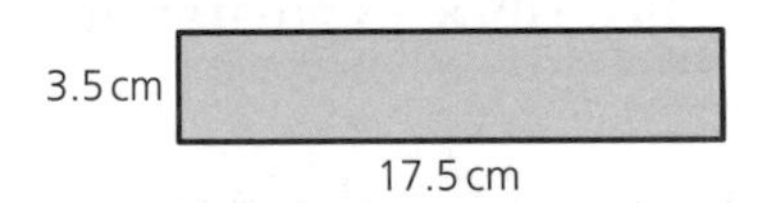

5 Tamsin usually mixes 400 ml of water with 100 ml of concentrate to make half a litre of orange juice. She has only 80 ml of concentrate. How much water should she use? __________ ml

6 An electric kettle normally costs £17.80

The price is reduced by 5%. What is the new price?

£__________

7 Tanya finished first in a downhill ski race with a time of 2 minutes and 58.5 seconds. Cristina finished 4.3 seconds behind Tanya. What was Cristina's time?

_______ minutes _______ seconds

8 What is $5 \times 11 \times 11$? __________

9 Connie buys 3 bottles of perfume costing £29.75 each. How much change does she receive from a £100 note?

£__________

10 What is a sixth of 312? __________

11 Gabriella is thinking of three integers (whole numbers) all less than 10. Their product is 70. What is the sum of the numbers? __________

12 Malcolm has the six number cards below.

What is the largest 5-digit odd number he can make by placing five of the cards side by side? __________

13 The Carroll diagram shows some information about a group of people.

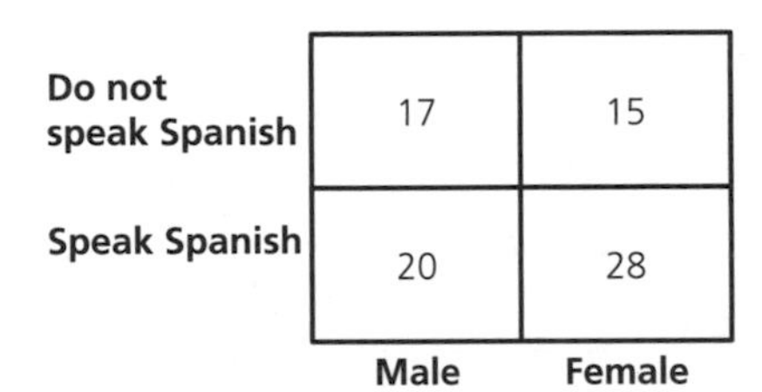

	Male	Female
Do not speak Spanish	17	15
Speak Spanish	20	28

What percentage of the group is Spanish-speaking males?

__________ %

14 What number is exactly half way between 45 and 103? __________

15 Which number between 200 and 209 is divisible exactly by 6? __________

16 What is the smallest number of British coins needed to pay exactly 98 pence? __________

17 Trisha has 45 marbles. A fifth are red and a third of the others are green. The rest are yellow. How many marbles are yellow?

____

18 How many of the following calculations give a result with a units digit 9?

____

7×17 $119 \div 7$ $999 - 99$ $65 + 104$

19 On 31 December, Jemima's age was recorded as 10 years 11 months. In which month is her birthday?

____

20 What is the sum of the numbers 1000 to 1003 inclusive?

____